MIX
Papier aus verantwortungsvollen Quellen
Paper from responsible sources
FSC® C105338

Kalpanadevi Tamilselvan

Metal Hydrazine Cinnamates

Synthesis and Characterization

Anchor Academic
Publishing

Tamilselvan, Kalpanadevi: Metal Hydrazine Cinnamates. Synthesis and
Characterization, Hamburg, Anchor Academic Publishing 2016

Buch-ISBN: 978-3-96067-034-6
PDF-eBook-ISBN: 978-3-96067-534-1
Druck/Herstellung: Anchor Academic Publishing, Hamburg, 2016

Bibliografische Information der Deutschen Nationalbibliothek:
Die Deutsche Nationalbibliothek verzeichnet diese Publikation in der Deutschen
Nationalbibliografie; detaillierte bibliografische Daten sind im Internet über
http://dnb.d-nb.de abrufbar.

Bibliographical Information of the German National Library:
The German National Library lists this publication in the German National Bibliography.
Detailed bibliographic data can be found at: http://dnb.d-nb.de

All rights reserved. This publication may not be reproduced, stored in a retrieval system
or transmitted, in any form or by any means, electronic, mechanical, photocopying,
recording or otherwise, without the prior permission of the publishers.

Das Werk einschließlich aller seiner Teile ist urheberrechtlich geschützt. Jede Verwertung
außerhalb der Grenzen des Urheberrechtsgesetzes ist ohne Zustimmung des Verlages
unzulässig und strafbar. Dies gilt insbesondere für Vervielfältigungen, Übersetzungen,
Mikroverfilmungen und die Einspeicherung und Bearbeitung in elektronischen Systemen.

Die Wiedergabe von Gebrauchsnamen, Handelsnamen, Warenbezeichnungen usw. in
diesem Werk berechtigt auch ohne besondere Kennzeichnung nicht zu der Annahme,
dass solche Namen im Sinne der Warenzeichen- und Markenschutz-Gesetzgebung als frei
zu betrachten wären und daher von jedermann benutzt werden dürften.

Die Informationen in diesem Werk wurden mit Sorgfalt erarbeitet. Dennoch können
Fehler nicht vollständig ausgeschlossen werden und die Diplomica Verlag GmbH, die
Autoren oder Übersetzer übernehmen keine juristische Verantwortung oder irgendeine
Haftung für evtl. verbliebene fehlerhafte Angaben und deren Folgen.

Alle Rechte vorbehalten

© Anchor Academic Publishing, Imprint der Diplomica Verlag GmbH
Hermannstal 119k, 22119 Hamburg
http://www.diplomica-verlag.de, Hamburg 2016
Printed in Germany

TABLE OF CONTENTS

CHAPTER I INTRODUCTION 1
  1.1 Introduction to coordination chemistry 1
    1.1.1. Coordination complex 2
    1.1.2. Coordination of ligand 2
    1.1.3. Coordination number of central metal atom / ion 3
    1.1.4. Coordination sphere and ionisation sphere 3
  1.2. Chemistry of Hydrazine 4
    1.2.1. Molecular structure of hydrazine 5
    1.2.2 Synthesis of hydrazine 5
    1.2.3. Infrared spectra of hydrazine, its salts and complexes 6
    1.2.4. Bond energy of hydrazine 8
    1.2.5. Applications of hydrazine 9
  1.3. Cinnamic acid 12
    1.3.1 Production 12
    1.3.2 Applications of cinnamic acid 12
  1.4. Chemistry of transition metals 14
    1.4.1. Cobalt 15
    1.4.2. Nickel 16
    1.4.3. Cadmium 17
    1.4.4. Zinc 18
    1.4.5. Iron 20
  1.5. Transition metal complexes of hydrazine 21
    1.5.1. Synthesis of transition metal hydrazine complexes 22
    1.5.2. Thermal reactivity 24
  1.6. Objectives of the work 27
  1.7. Organization of study 28

CHAPTER II EXPERIMENTAL DETAILS 29
  2.1. Materials 29
  2.2. Analytical methods 30
    2.2.1. Estimation of hydrazine in the complexes 30
    2.2.2 Estimation of metal ions in hetero bimetallic hydrazine cinnamates 30
    2.2.3. Estimation of metals in hetero trimetallic hydrazine cinnamates 31
  2.3. Physico-chemical techniques 32
    2.3.1 Infrared spectral analysis 32
    2.3.2. Thermogravimetric analysis (TGA) 33
    2.3.3. Scanning Electron Microscopy (SEM) – Energy Dispersive X-ray analysis (EDX) 33

CHAPTER III SYNTHESIS AND CHARACTERIZATION OF METAL HYDRAZINE CINNAMATES $[M(cin)_2(N_2H_4)_2]$ (M= Ni, Co, Zn or Cd) 35
   3.1. Synthesis and characterization of the complex $[Co(cin)_2(N_2H_4)_2]$ 35
      3.1.1. Synthesis of $[Co(cin)_2(N_2H_4)_2]$ 35
      3.1.2. Characterization of $[Co(cin)_2(N_2H_4)_2]$ 35
   3.2. Synthesis and characterization of the complex $[Ni(cin)_2(N_2H_4)_2]$ 39
      3.2.1. Synthesis of $[Ni(cin)_2(N_2H_4)_2]$ 39
      3.2.2. Characterization of $[Ni(cin)_2(N_2H_4)_2]$ 39
   3.3 Synthesis and characterization of $[Zn(cin)_2(N_2H_4)_2]$ 42
      3.3.1. Synthesis of $[Zn(cin)_2(N_2H_4)_2]$ 42
      3.3.2. Characterization of $[Zn(cin)_2(N_2H_4)_2]$ 43
   3.4. Synthesis and characterization of $[Cd(cin)_2(N_2H_4)_2]$ 46
      3.4.1. Synthesis of $[Cd(cin)_2(N_2H_4)_2]$ 46
      3.4.2. Characterization of $[Cd(cin)_2(N_2H_4)_2]$ 46
   3.5. Structure of the complexes $[M(cin)_2(N_2H_4)_2]$ (M=Co, Ni, Zn or Cd) 50

CHAPTER IV SYNTHESIS AND CHARACTERIZATION OF HETERO BIMETALLIC HYDRAZINE CINNAMATES $[MFe2(cin)3(N2H4)3]$ (M= Ni, Co, Zn or Cd) 51
   4.1. Synthesis and characterization of $[CoFe_2(cin)_3(N_2H_4)_3]$ 51
      4.1.1. Synthesis of $[CoFe_2(cin)_3(N_2H_4)_3]$ 51
      4.1.2. Characterization of $[CoFe_2(cin)_3(N_2H_4)_3]$ 51
   4.2. Synthesis and characterization of $[NiFe_2(cin)_3(N_2H_4)_3]$ 54
      4.2.1. Synthesis of $[NiFe_2(cin)_3(N_2H_4)_3]$ 54
      4.2.2. Characterization of $[NiFe_2(cin)_3(N_2H_4)_3]$ 55
   4.3. Synthesis and characterization of $[ZnFe_2(cin)_3(N_2H_4)_3]$ 58
      4.3.1. Synthesis of $[ZnFe_2(cin)_3(N_2H_4)_3]$ 58
      4.3.2. Characterization of $[ZnFe_2(cin)_3(N_2H_4)_3]$ 58
   4.4. Synthesis and characterization of $[CdFe_2(cin)_3(N_2H_4)_3]$ 62
      4.4.1. Synthesis of $[CdFe_2(cin)_3(N_2H_4)_3]$ 62
      4.4.2. Characterization of $[CdFe_2(cin)_3(N_2H_4)_3]$ 62

CHAPTER V SYNTHESIS AND CHARACTERIZATION OF HETERO TRIMETALLIC HYDRAZINE CINNAMATES 66
   5.1. Synthesis and characterization of $[Ni_{0.25}Co_{0.75}Fe_2(cin)_3(N_2H_4)_5]$ 66
      5.1.1. Synthesis of $[Ni_{0.25}Co_{0.75}Fe_2(cin)_3(N_2H_4)_5]$ 66
      5.1.2. Characterization of $[Ni_{0.25}Co_{0.75}Fe_2(cin)_3(N_2H_4)_5]$ 66
   5.2. Synthesis and characterization of $[Co_{0.8}Zn_{0.2}Fe_2(cin)_3(N_2H_4)_4]$ 70
      5.2.1. Synthesis of $[Co_{0.8}Zn_{0.2}Fe_2(cin)_3(N_2H_4)_4]$ 70
      5.2.2. Characterization of $[Co_{0.8}Zn_{0.2}Fe_2(cin)_3(N_2H_4)_4]$ 70

5.3. Synthesis and characterization of $[Ni_{0.8}Zn_{0.2}Fe_2(cin)_3(N_2H_4)_3]$ 74
    5.3.1. Synthesis of $[Ni_{0.8}Zn_{0.2}Fe_2(cin)_3(N_2H_4)_3]$ 74
    5.3.2. Characterization of $[Ni_{0.8}Zn_{0.2}Fe_2(cin)_3(N_2H_4)_3]$ 74
5.4. Synthesis and characterization of $[Cd_{0.3}Zn_{0.7}Fe_2(cin)_3(N_2H_4)_2]$ 77
    5.4.1. Synthesis of $[Cd_{0.3}Zn_{0.7}Fe_2(cin)_3(N_2H_4)_2]$ 77
    5.4.2. Characterization of $[Cd_{0.3}Zn_{0.7}Fe_2(cin)_3(N_2H_4)_2]$ 77

CHAPTER VI SUMMARY, CONCLUSIONS AND FURTHER SCOPE 82
  6.1 Summary of the work.. 82
  6.2 Conclusions... 83
  6.3 Further scope... 83

CHAPTER I
INTRODUCTION

Abstract

This chapter comprises of an introduction to the chemistry of coordination complexes. Some important aspects of hydrazine, cinnamic acid and transition metals are also detailed. Various methods of preparation of metal hydrazine carboxylates are written in brief. Objectives of the present study and organization of the thesis are also given.

1.1 Introduction to coordination chemistry

Coordination chemistry, the chemistry of metal complexes, is one of the most active research areas in inorganic chemistry. The study of coordination chemistry in the modern day context began with two notable scientists Alfred Werner and Sorphus Mads Jorgenson. The pioneering contribution of Werner to the study of coordination chemistry fetched him the Nobel Prize in Chemistry in 1913. Werner's basic ideas on the stereochemistry of metal complexes, mechanism of isomerisation etc., remain unchallenged even today despite all the advanced technical developments which have taken place since his days. However the advent of sophisticated physicochemical techniques of high precision and capability has considerably enriched our understanding of the nature of the metal-ligand bond, the structure and stereochemistry of metal complexes, their stability and other properties. Research has come long way from the time of Werner and Jorgenson, in terms of the growth that the coordination chemistry has experienced over the last few decades. Their work was a stepping stone for the development of modern inorganic chemistry which is truly a multidisciplinary one in the present day context.

Coordination chemistry encompasses such diverse fields as dyes, colour photography, mineral extraction, nuclear fuels, toxicology, bioinorganic chemistry, medicine, catalysis, material science, ceramics, microelectronics, photonics etc. Industries dealing with organic chemicals, pharmaceuticals, petrochemicals and plastics owe a lot to the findings in the field of coordination chemistry. Nature makes extensive use of coordination compounds and their study is becoming increasingly important in biology as well as in chemistry. Many of the biologically active compounds are complexes and even the simpler types of complexes have served as model compounds in investigating bodily process. The living system is partially supported by coordination compounds. Hemoglobin, an iron complex, carries oxygen to animal cells. Myoglobin, chlorophyll and cytochromes are some of the other important complex compounds in living systems. Inorganic compounds particularly metallic ions and complexes are essential cofactors in a variety of enzymes and proteins.

The elegance and the variety of the coordination compounds and the intriguing range of concepts that are required to interpret their behaviour have attracted many researchers to the study of their synthesis and to seek an understanding of their chemical reactions. The study of complexes has enabled the inorganic chemists to make significant progress in refining the concept of chemical bonding and to explain the influence that bonding has, on the various properties of the compounds.

1.1.1. Coordination complex

A coordination complex may be defined as a compound that result from the combination of two or more stable chemical species and retains its identity in the solid as well as dissolved state.

1.1.2. Coordination of ligand

Co-ordination complexes are formed by the union of a cation with one or more neutral or charged species (usually anions) which are attached with the central ion in complex compounds are called ligands or coordinating groups. The term ligand, which originated from the Latin word ligand originated from liogare, was first introduced by Alfred stock in 1896. According to Lewis, the ligands act as Lewis bases (Electron pair donors) and central metal ion acts as a Lewis acid (Electron acceptor), i.e. in most of the complexes, the ligands donate one or more electron pairs to the central metal ion.

$$M^{n+} + xL \rightarrow [ML_x]^{n+}$$

The ligands are attached to the central metal ion through their donor atoms. Ligands are classified according to the number of donor atoms contained and are known as uni, di, tri or quadridentate ligands, where the concept of teeth was introduced. This phenomenon of ring formation is called chelation. The term chelate was first introduced in 1920 by Morgan and Drew.

In some cases, multiple bonding will be there, i.e. ligands act simultaneously as donor and acceptor. For example, in metallic carbonyls, CO molecule can accept and donate election pairs.

1.1.3. Coordination number of central metal atom / ion

Co-ordination number of the central metal atom in a given complex compound is equal to the total number of donor atoms, which are actually attached with the central metallic atom. In other words, we can say that the co-ordination number of the central metallic atom is equal to the number of sites at which the ligands are attached to central metallic ion. In the case of complex compounds that contain only monodentate ligands, the coordination number of the central metallic atom is equal to the number of monodentate ligands coordinated to the metallic atom. This rule does not hold good for the complexes containing polydentate ligands.

Co-ordination number of metallic atom predicts the geometry of the complex compound. Thus for coordination number equal to 2,3,4,5 and 6, the geometry of the complex compound formed is linear, trigonal, planar, tetrahedral or square planar, trigonal bipyramidal and octahedral respectively. Coordination number gives us an idea about the way in which the ligands are arranged around the central metallic ion.

1.1.4. Coordination sphere and ionisation sphere

The central metal atom and ligand attached with it are always written in a square bracket called co-ordination or inner sphere. The portion outside the co-ordination sphere is called ionization or outer sphere. Thus in $[Co(NH_3)_5 Cl]Cl_2$, the square bracket which contains the central metal ion Co^{3+} and the ligand of five ammonia molecules and one chloride ion, in co-ordination sphere and the portion that contains two chloride ions is ionization sphere.

1.2. Chemistry of Hydrazine

Hydrazine is one of a series of compounds called hydronitrogens. It is an inorganic compound with the formula N_2H_4 and is also called diazane. It is a colourless flammable liquid with ammonia like odour. Hydrazine has physical properties very close to water, with a melting point of 2°C and a boiling point of 114°C. The similarity in its chemistry to water is as a result of strong intermolecular hydrogen bonding. The chemistry of hydrazine is appealing since it has a potent N-N bond, two electron pairs and four substitutable hydrogen atoms. It has a positive heat of formation ($\Delta H_f \approx$ 12 kcalmol^{-1}) and therefore is thermodynamically unstable. Practically, however, it is quite stable and can be handled safely.

Having two active nucleophilic nitrogens and four substitutable hydrogens, hydrazine is the starting material for many heterocyclics, where the rings contain one to four nitrogen atoms as well as other heteroatoms. Hydrazine, also called dinitrogen tetra hydride, dissolves in polar solvents such as water, alcohols, ammonia and amines. Hydrazine, the simplest diamine, with two lone pairs over nitrogen atoms is capable of forming variety of complexes with metal salts. The coordination of hydrazine to the metal ion in several complexes can either be unidentate or bridged bidentate. The bidentate coordination mode of hydrazine is common with transition metal ions containing inorganic and organic anions.

In acidic solution hydrazine exists in the form of either hydrazinium(1$^+$), ($N_2H_5^+$) or hydrazinium(2$^+$), ($N_2H_6^{2+}$) ion, among which the protonated form $N_2H_5^+$ is still capable of coordinating with metal ions. The $N_2H_6^{2+}$ ions are hydrolyzed by water in accordance with the equation,

$$N_2H_6^{2+} + H_2O \rightarrow N_2H_5^+ + H_3O^+$$

1.2.1. Molecular structure of hydrazine

$$H_2N-NH_2$$

Hydrazine can arise via a pair of ammonia molecules by removal of one hydrogen atom per molecule. Each H_2N-N subunit is pyramidal in shape. The N-N distance is 1.45 A^0 (145pm), and the molecule adopts a gauche conformation [1]. The rotational barrier is twice that of ethane. These structural properties resemble those of gaseous hydrogen peroxide, which adopts a "skewed" anticlinal conformation, and also experiences a strong rotational barrier.

1.2.2 Synthesis of hydrazine

Theodor Curtius was the first chemist who synthesized free hydrazine for the first time in 1889 by hydrolyzing diazoacetic ester with alkali [2].

$$2N_2CHCOOC_2H_5 + 4H_2O \longrightarrow 2N_2H_4 + (COOC_2H_5)_2 + (COOH)_2$$

Hydrazine was produced in the Olin Raschig [3] process from sodium hypochlorite (the active ingredient in many bleaches) and ammonia, a process announced in 1907. The reaction involves two stages.

$$NH_3 + NaOCl \longrightarrow NH_2Cl + NaOH \text{ (Fast)}$$
$$NH_3 + NH_2Cl + NaOH \longrightarrow N_2H_4 + NaCl + H_2O \text{ (Slow)}$$

In the Atofina–PCUK cycle [4,5], hydrazine is produced in several steps from acetone, ammonia, and hydrogen peroxide. Acetone and ammonia first react to give the imine followed by oxidation with hydrogen peroxide to the oxaziridine, a three-membered ring containing carbon, oxygen, and nitrogen, followed by ammonolysis to the hydrazone, a process that couples two nitrogen atoms. This hydrazone reacts with one more equivalent of acetone, and the resulting acetone azine is hydrolyzed to give hydrazine, regenerating acetone. Unlike the Raschig process, this process does not produce salt.

Alternative routes to synthesise hydrazine include the oxidation of urea with sodium hypochlorite [6].

$(H_2N)_2C=O + NaOCl + 2 NaOH \rightarrow N_2H_4 + H_2O + NaCl + Na_2CO_3$

All the processes do not yield anhydrous N_2H_4, but $N_2H_4.H_2O$. The majority of N_2H_4 is used as hydrate but several applications, in particular its use as a rocket propellant or as an explosive, require anhydrous hydrazine.

The water molecule can be removed from $N_2H_4.H_2O$ by chemical reaction with water binding chemicals, followed by distillation to separate the anhydrous N_2H_4 from the reaction product. The chemicals such as calcium carbide, sodiumhydroxide, calcium oxide and barium oxide or barium hydroxide can be used for this purpose.

Hydrazine remained as a laboratory curiosity for over 50 years. The bibliographic works on hydrazine by Audrieth and Ogg [7], Clark [8], Bottomley [9] and Schmidt [10] are indispensable bibles for hydrazine chemists.

The field of hydrazine chemistry and its applications are ever widening. Each year about 10,000 patents and articles on hydrazine derivatives are published. Hence another bibliographic work on hydrazine has to be undertaken.

1.2.3. Infrared spectra of hydrazine, its salts and complexes

One of the best features of an infrared spectrum is that the absorption or the lack of absorption in specific frequency regions can be correlated with specific stretching and bending motions and in some cases, with the relationship of these groups to the remainder of the molecule. IR spectra of hydrazine and its derivatives are studied in the finger print region between 1300 and 650 cm^{-1}. They have been reported for several hydrazine derivatives [11-21] and metal complexes [22, 23]. Normal coordinate analysis for N_2H_2, N_2H_4, $N_2H_5^+$ and $N_2H_6^{2+}$ has been carried out by Mielke and Ratajczak [24].

The special interest in the vibrational assignment of hydrazine is the N-N stretching frequency, since the presence of this frequency, has been used as a criterion for determining the mode of bonding of hydrazine to metal ions as well as to distinguish it from $N_2H_5^+$ and $N_2H_6^{2+}$ ions.

Braibanti *et al.* [22] have given a 'thumb rule' on the basis of earlier studies. In the complexes examined by them and also others γ_{N-N} could be found in the following frequency ranges.

N_2H_4 (in solid state)	875 cm^{-1}
N_2H_4 (unidentate)	930-940 cm^{-1}
N_2H_4 (bridging)	948-985 cm^{-1}
NH_2NHY (Y = COO, CSS)	986-1012 cm^{-1}
$N_2H_5^+$ cation (non-coordinated)	960-970 cm^{-1}
$N_2H_5^+$ cation (coordinated)	990-1015 cm^{-1}
$N_2H_6^{2+}$ cation	1020-1045 cm^{-1}

Hydrazine as a unidentate ligand, also shows N-N stretching at higher wave numbers, for example, 956 cm^{-1} in $Me_3B(N_2H_4)$, 952 cm^{-1} in $[Hg(N_2H_4)_2]Cl_2$ or 950 cm^{-1} in $SiF_4(N_2H_4)_2$.

Although the N-N stretchings for free $N_2H_5^+$ and bridging N_2H_4 overlap, they can be identified by fixing the molecular formulae by analytical and other techniques.

The assignment of the band at 875 cm^{-1} in the spectrum of hydrazine to γ_{N-N}, was questioned by Durig *et al.* [25], who assigned this band to an NH_2 rocking vibration and the band at 1126 cm^{-1} to γ_{N-N}. Satyanarayana and Nicholls [26] have recorded the infrared spectra and $M(N_2H_4)_2Cl_2$ (M = Mn, Fe, Co, Ni or Zn) complexes and assigned the absorption in the region 1150-1170 cm^{-1} to γ_{N-N} of bridged hydrazine. Despite these reservations, the frequency of γ_{N-N} is a useful indication of the type of coordinated hydrazine.

1.2.4. Bond energy of hydrazine

The bond energy of individual linkages that are holding the atoms together to a molecule determines the stability of a molecule. Obviously hydrazine is not a very stable molecule and some explanation for this behavior is sought by studying the bond energies of hydrazine and its decomposition products. Bond energies can be derived from the natural frequency of atoms oscillating in the molecular skeleton by extrapolation of spectroscopic data or from ionization and appearance potentials or from calorimetic data. Bond energies are identical with dissociation energies, required to break a particular bond. Table1 is a summary of bond dissociation energy data in hydrazine.

Table 1

Bond	Dissociation Energy		
	KJ/Mol	KCal Mol	E_v
$H-N_2H_3$	318±21	>6±2	33±2
$H-N_2H_3$	481	115	50
$H-N_2H_3$	326	78	34
$H-N_2H_3$	389	93	40
$H-N_2H_4$	368	88	-
H-NHNH	226±21	54±5	23±2
$H-N_2H$	314±21	75±5	33±2
$H-N_2$	-81±21	-21±5	-9±2
H_2N-NH_2	242±38	58±3	25±1
H_2N-NH_2	251±13	60±3	26±3
H_2N-NH_2	226	54	23
H_2N-NH_2	251	60	-

1.2.5. Applications of hydrazine

Hydrazine is a part of many organic syntheses, often those of practical significance in various fields.

➢ **Industrial uses**

Hydrazine is used in many processes including production of spandex fibers, as a polymerization catalyst, in fuel cells, solder fluxes and photographic developers, as a chain extender in urethane polymerizations and heat stabilizers. In addition, a semiconductor deposition technique using hydrazine has recently been demonstrated, with possible application to the manufacture of thin-film transistors used in liquid crystal displays.

Hydrazine is often used as an oxygen scavenger and corrosion inhibitor in boiler water treatment.

➢ **As a rocket fuel**

Hydrazine was first used as a rocket fuel during World War II for the Messerschmitt Me 163B (the first rocket-powered fighter plane), under the code name B-Stoff (hydrazine hydrate). When mixed with methanol (M-Stoff) and water it was called C-Stoff.

Hydrazine is also used as a low-power monopropellant for the maneuvering thrusters of spacecraft, and the Space Shuttle's auxiliary power units (APUs). In addition, monopropellant hydrazine-fueled rocket engines are often used in terminal descent of spacecraft. Such engines were used on the Viking program landers in the 1970s as well as the Phoenix lander and Curiosity rover who landed on Mars in May 2008 and August 2012, respectively.

In all hydrazine monopropellant engines, the hydrazine is passed by a catalyst such as iridium metal supported by high-surface-area alumina (aluminium oxide) or carbon nanofibers [27], or more recently molybdenum nitride on alumina [28], which causes it to decompose into ammonia, nitrogen gas and hydrogen gas according to the following reactions.

$$3N_2H_4 \rightarrow 4NH_3 + N_2$$
$$N_2H_4 \rightarrow N_2 + 2H_2$$
$$4NH_3 + N_2H_4 \rightarrow 3N_2 + 8H_2$$

First two reactions are extremely exothermic, the catalyst chamber can reach 800 °C in a matter of milliseconds and they produce large volumes of hot gas from a small volume of liquid making hydrazine a fairly efficient thruster propellant with a vacuum specific impulse of about 220 seconds. Third reaction is endothermic and so it reduces the temperature of the products, but also produces a greater number of molecules. The catalyst structure affects the proportion of the NH_3 that is dissociated in third reaction; a higher temperature is desirable for rocket thrusters, while more molecules are desirable when the reactions are intended to produce greater quantities of gas.

Other variants of hydrazine that are used as rocket fuel are MonoMethylHydrazine, $(CH_3)NH(NH_2)$ (also known as MMH) and Unsymmetrical DiMethylHydrazine, $(CH_3)_2N(NH_2)$ (also known as UDMH). These derivatives are used in two-component rocket fuels, often together with nitrogen tetroxide, N_2O_4, sometimes known as dinitrogen tetroxide. These reactions are extremely exothermic, and the burning is also hypergolic, which means that it starts without any external ignition source.

> **In fuel cells**

The Italian catalyst manufacturer Acta has proposed using hydrazine as an alternative to hydrogen in fuel cells. The chief benefit of using hydrazine is that it can produce 200mW/cm^2 more than a similar hydrogen cell without the need to use expensive platinum catalysts. As the fuel is liquid at room temperature, it can be handled and stored more easily than hydrogen. By storing the hydrazine in a tank full of a double-bonded carbon-oxygen carbonyl, the fuel reacts and forms a safe solid called hydrazone. By then flushing the tank with warm water, the liquid hydrazine hydrate is released. Hydrazine also has a higher electromotive force of 1.56 V compared to 1.23 V for hydrogen.

> **As a gun propellant**

A mixture of 63% hydrazine, 32% hydrazine hydrate and 5% water is a standard propellant for experimental bulk-loaded liquid propellant artillery. The propellant mixture above is notable for being one of the most predictable and stable mixtures, with a remarkably flat pressure profile during firing.

> **In nanoscience and nanotechnology**

Hydrazine being a fuel not only supports combustion but also lowers the decomposition temperature of the metal complexes. Coordination of hydrazine to the metal carboxylates markedly lowers the decomposition temperature of the metal carboxylates and yields nanosize metal oxides at comparatively lower temperatures. For this reason, hydrazinated carboxylates of metal as well as mixed metal carboxylates and dicarboxylates have been used for the synthesis of technologically important nano materials like γ-Fe_2O_3 [29-31], ferrites [32,33], manganites [34-37], cobaltites [38], etc. which can have several interesting electrical, magnetic, biological and catalytic properties.

> **Pharmaceutical applications**

Hydrazine derivatives have several potential pharmaceutical applications.

Though some hydrazine derivatives are proven and many others are suspected carcinogens, hydrazinium salts and some other hydrazine derivatives have been considered as pharmaceuticals for cancer treatment. Hydrazinium(III)sulphate was used to treat patients with tumors of the lung, breast, prostrate, ovary, lumph, cervix, thyroid and pancreas. Hydrazine derivatives of pyrimidine are claimed as cytostatic agents are said to be active against a number of malignant tumors [39]. Some antibiotics including the hydrazine group are Negamycin [40] and Rifmycin hydrazide [41]. Methisazone, an antiviral thiosemicarbazone marketed in Europe, can block replication of small pox virus in humans if administrated to contacts within 1-2 days after exposure and giving striking prophylactic protection against clinical small pox.

Hydrazine(1-hydrazinyl phthalazine) is used for treating blood pressure and hypertension. It lowers the blood pressure, especially diastolic pressure, while simultaneously increasing renal blood flow [42].

1.3. Cinnamic acid

Cinnamic acid, an unsaturated carboxylic acid of the formula $C_6H_5CHCHCO_2H$, is a white crystalline compound slightly soluble in water and freely soluble in many organic solvents.

Cinnamic acid (3–phenyl–2–propanoic acid) being a derivative of phenylalanine comprises a relatively large family of organic isomers [43]. It occurs both as cis and trans forms, the latter being the most important.

Cis   trans

1.3.1 Production

The original synthesis of cinnamic acid involves the Perkin reaction, which entails the base-catalysed condensation of acetic anhydride and benzaldehyde.

1.3.2 Applications of cinnamic acid

Cinnamic acid is used in flavors, synthetic indigo and certain pharmaceuticals. A major use is in the manufacturing of the methyl, ethyl, and benzyl esters for the perfume industry [44]. Cinnamic acid is a precursor to the sweetener aspartame via enzyme-catalysed amination to phenylalanine [45]. It is also used in food additive, cosmetic and pesticide industries. Cinnamic acid is also a kind of self-inhibitor produced by fungal spore to prevent germination. *trans-* Cinnamic acid has a long history of human use as a component of plant-derived scents and flavourings [46]. It belongs to the class of auxin, which is recognized as plant hormones regulating cell growth and differentiation [47]. It has antibacterial, anti fungal and anti parasitic properties and its derivatives are important pharmaceuticals for high blood pressure and stroke prevention.

Commercial cinnamic acid, a phenylacrylic acid structure compound, is used in converting to its esters such as methyl, ethyl, and benzyl cinnamate for the perfume and flavour application. Cinnamates can act as optical filters or deactivate substrate molecules that have been excited by light for the protection polymers and organic substances. They are used as sunscreen agents to reduce skin damage by blocking UV-A, B.

Cinnamic acid and its phenolic analogues are natural substances. Chemically, in cinnamic acid, the 3-phenyl acrylic acid functionality offers three main reactive sites: substitution at the phenyl ring, addition at the α,β- unsaturation and the reactions of the carboxylic acid functionality. Owing to these chemical aspects, cinnamic acid derivatives received much attention in medicinal research as traditional as well as recent synthetic antitumor agents.

In spite of their rich medicinal tradition, cinnamic acid derivatives and their anticancer potentials remained underutilized for several decades since the first published clinical use in 1905. In recent years, *trans* cinnamic acid derivatives have also attracted much attention due to their antioxidative [48], antitumor [49], anticancer [50] and antimicrobial [51,52] properties. Cinnamic acid and its derivatives have a century-old history as antituberculosis agents. Cinnamic acid possesses an α,β- unsaturated carbonyl moiety, which can be considered as a Michael acceptor, an active moiety often employed in design of anti-cancer drugs [53].

1.4. Chemistry of transition metals

In chemistry, the term transition metal, sometimes also called transition element has two possible meanings:
- The IUPAC definition states that a transition metal is an element whose atom has an incomplete d sub-shell.
- Most scientists describe a "transition metal" as any element in the d-block of the periodic table, which includes groups 3 to 12 on the periodic table. All elements in the d-block are metals.

Jensen [54] has reviewed the history of the terms transition element (or metal) and d-block. The word *transition* was first used to describe the elements now known as the d-block by the English chemist Charles Bury in 1921, who referred to a transition series of elements during the change of an inner layer of electrons (for example n=3 in the 4th row of the periodic table) from a stable group of 8 to one of 18, or from 18 to 32 [55].

The elements of d-block and f-block are collectively known as transition elements. A small distinction made between d-bock and f-block elements, calling f-block elements as inner transition elements and d-block is outer transition or transition elements.

The transition metals are well known for their complex formation tendency. Factors which favour the transition metals to form stable complexes may be due to the following.

I) Comparitively small size of the ions and high ionic charge

Both these factors enhance the attraction of the caution for a negative ion or a polar molecule and so favour complex formation. For example, the complex ions of Cr^{3+}, Fe^{3+} and Co^{3+} are more numerous and more stable than those of Cr^{2+}, Fe^{2+} and Co^{2+}.

II) The availability of (n-1)d orbitals

The (n-1) d orbital can be used to accept electron pairs from the complexing ligands. Even if the transition metal ion in its ground state does not possess vacant d-orbital, these may be often made available by rearrangement of the electrons in these orbitals. This process involves additional pairing of (n-1) d electrons.

The properties and applications of some important transition metals are described below.

1.4.1. Cobalt

Cobalt is a ferromagnetic metal with a specific gravity of 8.9. Pure cobalt is not found in nature, but compounds of cobalt are common. Small amounts of it are found in most rocks, soil, plants, and animals. The Curie temperature is 1115 °C [56] and the magnetic moment is 1.6–1.7 Bohr magnetons per atom [57]. In nature, it is frequently associated with nickel, and both are characteristic minor components of meteoric iron. Cobalt has a relative permeability two thirds that of iron [58]. Metallic cobalt occurs as two crystallographic structures: hcp and fcc. The ideal transition temperature between the hcp and fcc structures is 450 °C, but in practice, the energy difference is so small that random intergrowth of the two is common [59-61].

Cobalt is a weakly reducing metal that is protected from oxidation by a passivating oxide film. It is attacked by halogens and sulfur. Heating in oxygen produces Co_3O_4 which loses oxygen at 900 °C to give the monoxide CoO [62]. The metal reacts with F_2 at 520 K to give CoF_3, with Cl_2, Br_2 and I_2, the corresponding binary halides were formed. It has no reaction with H_2 and N_2 even when heated, but it does react with boron, carbon, phosphorus, arsenic and sulphur [63]. At ordinary temperatures, it reacts slowly with mineral acids, and very slowly with moist and air.

1.4.1.1. Oxidation state

Common oxidation states of cobalt include +2 and +3, although compounds with oxidation states ranging from −3 to +4 are also known. A common oxidation state for simple compounds is +2.

1.4.1.2. Applications

> **Industrial and medicinal field**

Cobalt-based super alloys consume most of the produced cobalt [64,65]. These alloys are corrosion and wear-resistant. The stability of these alloys makes them suitable for use in turbine blades for gas turbines and jet aircraft engines. Special cobalt-chromium-molybdenum alloys like Vitallium are used for prosthetic parts such as hip and knee replacements [66]. The special alloys of aluminium, nickel, cobalt and iron, known as Alnico, and of samarium and cobalt (samarium-cobalt magnet) are used in permanent magnets [67]. It is also alloyed with 95% platinum for jewellery purposes.

Lithium cobalt oxide (LiCoO$_2$) is widely used in lithium ion batteries [68]. Nickel-cadmium [69] (NiCd) and nickel metal hydride [70] (NiMH) batteries also contain significant amounts of cobalt; the cobalt improves the oxidation capabilities of nickel in the battery [69].

> **As a catalyst**

Several cobalt compounds are used in chemical reactions as oxidation catalysts. Cobalt acetate is used for the conversion of xylene to terephthalic acid, the precursor to the bulk polymer polyethylene terephthalate. Cobalt carboxylates catalysts (known as cobalt soaps) are also used in paints, varnishes and inks as "drying agents"[68]. Cobalt-based catalysts are also important in reactions involving carbon monoxide. It is used in steam reforming, in hydrogen production. Cobalt is also a catalyst in the Fischer–Tropsch process [71]. Hydrodesulfurization of petroleum uses a catalyst derived from cobalt and molybdenum. This process helps to rid petroleum of sulfur impurities that interfere with the refining of liquid fuels [68].

> **As a pigment and coloring compound**

Before the 19th century, the predominant use of cobalt was as pigment. Since the Middle Ages, it has been involved in the production of smalt, a blue colored glass. Smalt was widely used for the coloration of glass and as pigment for paintings [72]. In 1780, Sven Rinman discovered cobalt green and in 1802 Louis Jacques Thénard discovered cobalt blue [73]. These two variety are used as pigments for paintings because of their superior stability [74,75]

1.4.2. Nickel

Nickel is an element with the chemical symbol Ni and atomic number 28. It is a silvery-white lustrous metal with a slight golden tinge. It is hard and ductile. It is magnetic at room temperature or near room temperature. Its Curie temperature is 355 °C, meaning that bulk nickel is non-magnetic above this temperature [76]. The unit cell of nickel is a face centered cubic with the lattice parameter of 0.352 nm.

1.4.2.1. Oxidation state

The most common oxidation state of nickel is +2, but compounds of Ni^0, Ni$^+$ and Ni^{3+} are well known, and Ni^{4+} also been demonstrated [77].

1.4.2.2. Applications

The fraction of global nickel production presently used for various applications is as follows: 46% for making nickel steels; 34% in nonferrous alloys and superalloys; 14% for electroplating, and 6% for other uses [78,79].

Nickel is used in many specific and recognizable industrial and consumer products, including stainless steel, alnico magnets, coinage, rechargeable batteries, electric guitar strings, microphone capsules, and special alloys. It is used for plating and as a green tint in glass. It is widely used in many other alloys, such as nickel brasses and bronzes and alloys with copper, chromium, aluminium, lead, cobalt, silver and gold (Inconel, Incoloy, Monel, Nimonic) [80]. Alnico nickel alloy is used to make "horseshoe magnet". Nickel foam or nickel mesh is used in gas diffusion electrodes for alkaline fuel cells [81,82]. Nickel and its alloys are frequently used as catalysts for hydrogenation reactions. Raney nickel, a finely divided nickel-aluminium alloy, is commonly used as 'Raney-type' catalysts.

It is a naturally magnetostrictive material [83]. It is also used as a binder in the cemented tungsten carbide or hardmetal industry. It can make the tungsten carbide magnetic and adds corrosion-resistant properties to the cemented tungsten carbide [84]

1.4.2.3. Biological role of nickel

Nickel plays several important roles in the biology of microorganisms and plants [85]. The plant enzyme urease (an enzyme that assists in the hydrolysis of urea) contains nickel. The NiFe-hydrogenases also contain nickel in addition to iron-sulfur clusters. A nickel-tetrapyrrole coenzyme, Cofactor F430, is present in the methyl coenzyme M reductase, which powers methanogenic archaea.

1.4.3. Cadmium

Cadmium is an element with the symbol Cd and atomic number 48. It occurs as a minor component in most zinc ores and therefore is a by-product of zinc production. It is a soft, malleable, ductile, bluish-white divalent metal. It is similar in many respects to zinc but forms coordination compounds [86]. Unlike other metals, cadmium is resistant to corrosion and as a result it is used as a protective layer when deposited on other metals. As a bulk metal, it is insoluble in water and is non inflammable [87].

1.4.3.1. Oxidation state

Cadmium usually has an oxidation state of +2; it also exists in the +1 state.

1.4.3.2. Applications

Cadmium pigment is commonly used in electroplating [88]. Cadmium electroplating can be found in the aircraft industry due to the ability to resist corrosion when applied to steel components [88].

Cadmium is used as a key component in battery production. In 2009, 86% of cadmium was used in batteries, predominantly in rechargeable nickel-cadmium batteries. Nickel-cadmium cells have a nominal cell potential of 1.2 V.

1.4.4. Zinc

It has the symbol Zn and atomic number 30. It is also referred in nonscientific contexts as spelter [89]. It is a bluish-white, lustrous, diamagnetic metal. It has a hexagonal crystal structure [90]. It is hard and brittle at most temperatures but becomes malleable between 100 and 150 °C [91]. Above 210 °C, the metal becomes brittle again and can be pulverized by beating. It is a fair conductor of electricity. It has relatively low melting (419.5 °C, 787.1 F) and boiling points (907 °C) [92]. Its melting point is the lowest of all the transition metals aside from mercury and cadmium [92].

1.4.4.1. Oxidation state

The chemistry of zinc is dominated by the +2 oxidation state. Zn compounds with +2 oxidation state are formed by the loss of outer shell *s* electrons, which yields a zinc ion with the electronic configuration $[Ar]3d^{10}$ [93]. In aqueous solution an octahedral complex, $[Zn(H_2O)_6]^{2+}$ is the predominant species [94]. The volatilization of zinc in combination with zinc chloride at temperature above 285 °C forms of Zn_2Cl_2, a zinc compound with +1 oxidation state [95]. No compounds of zinc in oxidation states other than +1 or +2 are known [96].

1.4.4.2. Applications

Major applications of zinc include (numbers are given for the US) [97]

                    Galvanizing (55%)

                    Alloys (21%)

                    Brass and bronze (16%)

                    Miscellaneous (8%)

➤ **As an anti-corrosion agent and in batteries**

Zinc is most commonly used as an anti-corrosion agent. Galvanization, which is the coating of iron or steel to protect the metals against corrosion, is the most familiar form of using zinc. In 2009 in the United States, 55% or 893 thousand tonnes of the zinc metal were used for galvanization [97]. It is applied electrochemically or as molten zinc by hot-dip galvanizing or spraying. Galvanization is used on chain-link fencing, guard rails, suspension bridges, lightposts, metal roofs, heat exchangers and car bodies.

It is also used to cathodically protect metals that are exposed to sea water from corrosion [98]. A zinc disc attached to a ship's iron rudder will slowly corrode while the rudder stays unattacked. With a standard electrode potential (SEP) of −0.76 volts, it is used as an anode material for batteries. Powdered zinc is used in alkaline batteries and sheets of zinc metal act as anodes in zinc–carbon batteries [99,100]. It is used as the anode or fuel of the zinc-air battery/fuel cell [101-103].

➤ **In industry**

A widely used alloy which contains zinc is brass, depending upon the type of brass, copper is alloyed with 3% to 45% zinc. Brass is generally more ductile and stronger than copper and has superior corrosion resistance. These properties make it useful in communication equipment, hardware, musical instruments, and water valves.

Alloys of zinc with small amounts of copper, aluminium, and magnesium are useful in die casting as well as spin casting, especially in the automotive, electrical and hardware industries. These alloys are marketed under the name Zamak [104], eg., zinc aluminium. Another alloy, marketed under the brand name Prestal, contains 78% zinc and 22% aluminium and is reported to be nearly as strong as steel but as malleable as plastic [105]. This superplasticity of the alloy allows it to be molded using die casts made of ceramics and cement.

Zinc oxide is used as a white pigment in paints. Zinc chloride is often added to lumber as a fire retardant and it is used as a wood preservative [106]. Zinc sulfide (ZnS) is used in luminescent pigments such as on the hands of clocks, X-ray and television screens, and luminous paints. Crystals of ZnS are used in lasers that operate in the mid-infrared part of the spectrum [107]. Zinc sulfate is used in dyes and pigments. Zinc pyrithione is used in antifouling paints [108].

Zinc powder is used as a propellant in model rockets [109]. When a compressed mixture of 70% zinc and 30% sulfur powder is ignited there is a violent chemical reaction [109]. This produces zinc sulfide, together with large amounts of hot gas, heat and light.

1.4.4.3. Biological role of Zinc

It possess antioxidant properties, which may protect against accelerated aging of the skin and muscles of the body [110]. It also helps speed up the healing process of an injury [110]. It is also beneficial to the body's immune system.

Zinc lactate is used in toothpaste to prevent halitosis [111]. Zinc pyrithione is widely applied in shampoos because of its anti-dandruff function [112]. Zinc ions are effective antimicrobial agents even at low concentrations [113].

1.4.5. Iron

Iron is an element with the symbol Fe (from Latin: *ferrum*) and atomic number 26. Molten iron cools down it crystallizes at 1538 °C into its δ allotrope, which has a body-centered cubic (bcc) crystal structure. As it cools further its crystal structure changes to face-centered cubic (fcc) at 1394 °C. At 912 °C the crystal structure again becomes bcc as α-iron or ferrite is formed, and at 770 °C (the Curie point, T_c) iron becomes magnetic. As the iron passes through the Curie temperature there is no change in crystalline structure, but there is a change in "domain structure", where each domain contains iron atoms with a particular electronic spin.

1.4.5.1. Oxidation state

Iron forms compounds mainly in the +2 and +3 oxidation states. Traditionally, iron(II) compounds are called ferrous, and iron(III) compounds ferric. It also occurs in higher oxidation states, for example the purple potassium ferrate (K_2FeO_4) contains iron in +6 oxidation state. Iron(IV) is a common intermediate in many biochemical oxidation reactions [114,115]. Numerous organometallic compounds contain formal oxidation states of +1, 0, −1, or even −2. The oxidation states and other bonding properties are often assessed using the technique of Mössbauer spectroscopy [116]. There are also many mixed valence compounds contain both iron(II) and iron(III) centers, for example Prussian blue [$Fe_4(Fe(CN)_6)_3$] [115].

1.4.5.2. Applications

Iron is the most widely used of all the metals, accounting for 95% of worldwide metal production. Its low cost and high strength in engineering applications such as the construction of machinery and machine tools, automobiles, the hulls of large ships and structural components for buildings. Since pure iron is quite soft, it is most commonly used in the form of steel. Commercially available iron is classified based on purity and the abundance of additives. Pig iron has 3.5–4.5% carbon [117] and contains varying amounts of contaminants such as sulfur, silicon and phosphorus. Pig iron is not a saleable product, but rather an intermediate step in the production of cast iron and steel from iron ore. Cast iron contains 2–4% carbon, 1–6% silicon, and small amounts of manganese.

Iron(III) chloride is used in water purification, sewage treatment, in the dyeing of cloth, as a coloring agent in paints, as an additive in animal feed and as an etchant for copper in the manufacture of printed circuit boards [118]. Iron (II) sulfate is used as a precursor to other iron compounds.

1.4.5.3. Biological role of Iron

Iron is abundant in biology. Iron-proteins are found in all living organisms, ranging from the evolutionarily primitive archaea to humans. The most commonly known and studied "bioinorganic" compounds of iron are the heme proteins: examples are hemoglobin, myoglobin and cytochrome P450. These compounds can transport oxygen, store oxygen and transfer electrons. Metalloproteins are a group of proteins with metal ion cofactors. Some examples of iron metalloproteins are ferredoxin and rubredoxin. Many enzymes vital to life contain iron, such as catalase, lipoxygenases, and IRE-BP.

1.5. Transition metal complexes of hydrazine

Hydrazine acts as a ligand to form metal complexes. Its very good liganding property arises from the presence of two free lone pair of electrons. The possible coordination modes of hydrazine molecule are monodentate, bridging and chelating bidentate. Among these, only a few reports on metal complexes with hydrazine in chelating bidentate fashion are available in the literature [119]. But a large number of metal complexes in which hydrazine acting as monodentate [120-122] as well as bidentate bridging ligand [123-126] has been synthesized and characterized. The mono protonated hydrazine i.e., hydrazinium(1+) cation that retains a lone pair of electron can also act as a ligand and metal complexes of this cationic ligand [127-129] have also been reported in the literature.

The methyl, ethyl, phenyl and dimethyl substituted hydrazines also act as ligands. However, there is a reduction in basicity with increasing alkylation of hydrazines, as measured by the acid dissociation constants [130]. The complexes containing bridging phenyl hydrazines have not been prepared probably for steric reasons [131,132]. Most alkyl and aryl-substituted hydrazines act only as monodentate ligands. However, because of more steric hindrance from phenyl and the substituents on it, the coordination numbers are generally lower than that with alkyl hydrazines.

From complexometric titration, Bisacchi and Goldwhite [133] have shown that the donor abilities of the hydrazines are in the order:

$$N_2H_4 > CH_3NHNH_2 > C_2H_5NHNH_2 > (CH_3)_2NNH_2$$

1.5.1. Synthesis of transition metal hydrazine complexes
1.5.1.1. Reactions of hydrazine and its salts with metal

The high dielectric constant of anhydrous hydrazine suggests that it could be a moderate solvent for many ionic compounds. It is not altogether unexpected to find that hydrazine salts when dissolved in hydrazine or hydrazine hydrate behave as acids. Thus, metals like Mg, Fe, Co, Ni, Zn or Cd dissolved in a solution containing hydrazine hydrate and hydrazinium or ammonium salts liberate hydrogen [134-136].

$$M + 2N_2H_5X \xrightarrow{N_2H_4, H_2O} M(N_2H_4)_2 X_2 + H_2$$

where $X = 0.5SO_4$, $0.5C_2O_4$, N_3, ClO_4 etc. Some mixed metal oxalate derivatives such as $MFe_2(C_2O_4)_3(N_2H_4)_x$ [137] (M = Mg, Mn, Co, Ni or Zn; x = 5 or 6) and $MgFe_2(N_2O_2)_3(N_2H_4)_5$ [138] have been synthesized using the above procedure. The complex $(N_2H_5)_2Mg(SO_4)_2$ has been prepared by the reaction of magnesium powder and ammonium sulphate in the presence of hydrazine hydrate [139].

1.5.1.2. Reactions of hydrazine with metal salts

The insoluble complexes $M(N_2H_4)_2X_2$ [140-149], (M = Mn, Fe, Co, Ni, Zn or Cd and X = Cl, Br, I, $0.5SO_4$, NCS, HCOO, CH_3COO, $0.5C_2O_4$, H_2NCH_2COO, $HOCH_2COO$, etc.,) are the usual products of reaction between excess hydrazine hydrate and first row transition metal salts. The trishydrazine complexes $M(N_2H_4)_3X_2$ [150-152], ($X = 0.5\ SO_4$, $0.5SO_3$, $0.5S_2O_3$, NO_3, etc.,) have been prepared by the reaction between the transition metal salts and hydrazine hydrate. The trishydrazine metal glycinates and glycolates $M(XCH_2COO)_2(N_2H_4)_3$, ($X = NH_2$ or OH and $M = Mn, Co, Ni, Zn$ or Cd) have been prepared [153] by mixing the metal nitrate hydrates and a mixture of the acid and excess hydrazine hydrate. Divalent Co, Ni, Zn and Cd pyrazine carboxylate hydrazinates of the formula $M(pyzCOO)_2.nN_2H_4.xH_2O$ and $Mpyz(COO)_2.N_2H_4.xH_2O$ have been obtained by the reaction of respective metal nitrate hydrates with 2-pyrazine carboxylic (HpzCOO)/2,3-pyrazinecarboxylic carboxylic ($H_2pyz(COO)_2$) acid and hydrazine hydrate [154].

1.5.1.3. Reactions of hydrazine and metal salt in the presence of CO_2

The hydrazine carboxylate $N_2H_3COO^-$ anion plays an important role in the preparation of low temperature precursors to metal or mixed metal oxides. This anion containing complexes have been prepared by passing CO_2 into an aqueous solution containing the metal salt and hydrazine hydrate till the solution is saturated with CO_2. The clear solution on slow evaporation at room temperature produces the crystals of the complexes. Thus the complexes $M(N_2H_4)_2(N_2H_3COO)_2$ ($M = Co, Ni$ or Zn), $M (N_2H_3COO)_2$ ($M = Mn, Co, Ni, Cu$ or Zn), $N_2H_5M(N_2H_3COO)_3$. H_2O ($M = Fe, Co, Ni$ or Zn) and $N_2H_5Sc(N_2H_3COO)_4.3H_2O$ have been prepared [155-165] by this method. The complexes $M(N_2H_3COO)_2.2H_2O$ ($M = Co$ or Ni); $Co(N_2H_3COO)_2$, $Co(N_2H_3COO)_2N_2H_4$, $Ni(N_2H_3COO)_2N_2H_4.H_2O$ and $Nd(N_2H_3COO)_3.3H_2O$ have also been prepared in a similar way by Macek and Rahten [166]. $Mg(N_2H_3COO)_2.H_2O$, $Ca(N_2H_3COO)_2.H_2O$, $Zn(N_2H_3COO)_2$ and $Cu(N_2H_3COO)_20.5H_2O$ have been synthesized by the same method [164-167]. Similar type of complexes has also been studied by Patil et al. [168]. The crystal structures of $Cd(N_2H_3COO)_2.H_2O$ [169] and $Mn(N_2H_3COO)_2.2H_2O$ [170] have been studied to ascertain the coordination behaviour of hydrazinecarboxylate anion. The complexes $Cu(N_2H_3COO)_2.H_2O$, $Cr(N_2H_3COO)_2.H_2O$ and $Cr(N_2H_3COO)_3.3H_2O$ have also been prepared by the reaction between metal salts and N_2H_3COOH and studied [171, 172]. The hydrazinecarboxylate derivatives of rare earth metal ions $Ln(N_2H_3COO)_3$(Ln =

La, Ce, Pr, Nd or Sm) and Ln(N$_2$H$_3$COO)$_3$(H$_2$O)$_3$ (Ln= La or Nd) have been prepared and characterized by Sivasankar et al. [173].

1.5.1.4. Reactions of hydrazine carboxylate complexes with acids

Whenever it is not possible to prepare hydrazinium metal complexes with particular anion, the same can be prepared conveniently by decomposing hydrazinium metal hydrazinecarboxylates with dilute acids of the corresponding anion. Example (N$_2$H$_5$)$_2$M(NCS)$_4$. 2H$_2$O (M = Co or Ni).

Complexes have been prepared by Kumar *et al.*, [174] by adding freshly prepared solid N$_2$H$_5$M(N$_2$H$_3$COO)$_3$.H$_2$O to dilute thiocyanic acid in small portions while maintaining the reaction temperature around 0 °C. The (N$_2$H$_5$)$_2$MnF$_4$, (N$_2$H$_5$)$_2$MCl$_4$.2H$_2$O (M = Co or Ni) and N$_2$H$_5$UO$_2$ (CH$_3$COO)$_3$ complexes have been prepared by the same procedure [175].

In spite of a number of methods described for the preparation of the complexes, it is not possible to detail all the possibilities as it is still a growing field. For example, some complexes have been prepared in non-aqueous medium and (N$_2$H$_5$)$_2$UF$_6$ has been prepared by the reaction between UF$_6$ and N$_2$H$_5$F in anhydrous hydrazine [176]. The lanthanide hydrazine complexes with anions like halides, carbonate, nitrate, sulphate, perchlorate, acetate and oxalate [10] and uranyl hydrazine complexes with chloride and sulphate [151] have been prepared by the addition of hydrazine hydrate to the metal salts in an aqueous or alcoholic medium. Brzyska and Goral [177] have prepared the complexes Ln$_2$ (C$_8$H$_4$O$_4$)$_3$ (N$_2$H$_4$)$_4$ 2C$_8$H$_6$O$_4$ (Ln = La or Ce) and Nd$_2$ (C$_8$H$_4$O$_4$)$_3$ (N$_2$H$_4$)$_3$ by crystallization from 1:6:6 molar ratio mixtures of lanthanide salt, hydrazine hydrate and phthalic anhydride.

1.5.2. Thermal reactivity

Thermal reactivity of the complexes varies from explosion → deflagration → decomposition depending upon the anion. Transition metal perchlorate, nitrate and azide hydrazines are primary explosives, non-transition metal (Li$^+$, Mg^{2+}, Al^{3+}) perchlorate, nitrate and azide hydrazines and transition metal oxalate, sulphite and hydrazinecarboxylate hydrazine complexes deflagrate and the rest simply decompose with the loss of hydrazine. The deflagrating nature of metal hydrazines has been used in the preparation of ferrites [178,179] and cobaltites [180,181]. It is rather surprising that thermolysis of $Mg(N_3)_2(N_2H_4)_2$ gave a blue coloured residue which showed a strong IR absorption at 2100 cm^{-1} characteristic of molecular nitrogen. The composition of the residue has been fixed as $Mg(NH_2)_2N_2$ by chemical analysis and TG studies [136]. The trishydrazine complexes are considerably less stable both thermally and in air than the corresponding bis-hydrazine complexes. The complexes $M(XCH_2COO)_2(N_2H_4)_3$, (X = NH_2 or OH and M = Mn, Co, Ni, Zn or Cd) decompose violently above 200°C in an exothermic single-step to form metal powders [153]. Thus, thermal properties of the complexes differ depending on the composition, the metal ion, type of the anion, the coordination mode of hydrazine and the atmosphere used during the experiments.

1.5.2.1. Thermal decomposition of metal hydrazine complexes

Thermal decomposition of metal hydrazine complexes with a variety of anions such as halides, $0.5SO_4$ [142], ClO_4, NO_3 and N_3 [134] have been studied. Depending upon the anion, the decomposition path changes dramatically. For example, the chlorate, perchlorate, nitrate and azide hydrazine complexes on heating decompose exothermically (violently) giving mostly metal oxides as the final residue, whereas hydrazine complexes, $M(N_2H_4)_nX_2$ [144, 182-187], (n = 2 or 3; X = F, Cl, Br, NCS, $0.5SO_4$, $0.5C_2O_4$, etc.) decompose exothermically through an intermediate, $M(N_2H_4)X_2$, to MX_2, MOX_2, MO, M_2O_3 or M. Thermal reactivity of $MFe_2(C_2O_4)_3$ $(N_2H_4)_x$ [137], (M = Mg, Co, Ni or Zn and X =5 or 6) and MFe_2 $(N_2O_2)_3(N_2H_4)_5$ [138] has been reported and these complexes decompose at low temperature to give ferrites as the final product. Preparation and thermal reactivity of $MgC_2O_4(N_2H_4)_2$ and $Mg(N_3)_2$ $(N_2H_4)_2$ [136] have also been reported.

Thermal decomposition of metal carboxylate hydrazines is more interesting due to their easier combustibility. For example, metal hydrazine formate [148], acetate [149], chloroacetate, glycinate and glycolate [188], malonate and succinate [189,190] complexes have been studied by simultaneous TG-DTA analysis. These complexes have been reported to decompose at lower temperatures than their non-carboxylate counterparts. Moreover, the oxalate complexes exhibit autocatalytic decomposition. This behaviour has been attributed to the simultaneous exothermic decomposition of hydrazine and metal salt [133]. This phenomenon has been made use of in the preparation of nickel ferrite nanoparticles [191] and cobaltites [181, 192, 193] by the low temperature decomposition of $NiFe_2(C_4H_2O_4)_3 6N_2H_4$ [177] and $M_{1/3}Co_{2/3}(C_2O_4)(N_2H_4)_2$, (M = Mg or Ni), respectively. Large surface area CeO_2 has been prepared by the thermal decomposition of cerium oxalate hydrazine complex [194]. The decomposition of nickel hydrazine glycinate and glycolate complexes [195] has been reported to be violently exothermic and lead to explosion if the samples are heated in bulk. They give metal powder as the final product, even in air, which is something unusual.

Metal hydrazinecarboxylates decompose in air at a low temperature (75-200°C) to yield fine particle oxide materials. Thermal studies on $M(N_2H_3COO)_2.nH_2O$, (M = Ca, Mg, Mn, Fe, Co, Ni, Zn or Cu; n = 0, 0.5, 1, 2, 3); and $M(N_2H_3COO)_2 (N_2H_4)_2$, (M = Mn, Fe, Co, Ni or Zn) have been carried out extensively in air or inert atmosphere [196-199]. The thermal property of $Nd(N_2H_3COO)_3.3H_2O$ in an inert atmosphere [167], the synthesis of $La(N_2H_3COO)_3.2H_2O$ [200] and the thermal reactivity of $Ln(N_2H_3COO)_3(Ln = La, Ce, Pr, Nd$ or Sm) and $Ln(N_2H_3COO)_3(H_2O)_3$ (Ln= La or Nd) [173] have been reported already. The decomposition is autocatalytic and accompanied by swelling due to the evolution of large amounts of gases like NH_3, H_2O, H_2 and CO_2 [201]. The preparation of γ-Fe_2O_3 and Co-doped γ-Fe_2O_3, the commonly used recording materials, has been achieved by the thermal decomposition of iron hydrazine carboxylates in a single step [202]. Similarly ultrafine ferrites (MFe_2O_4, $Ni_xZn_{1-x}Fe_2O_4$ and $Mn_xZn_{1-x}Fe_2O_4$) and fine particle cobaltites (MCo_2O_4) have been obtained at a very low temperature (300°C) by the thermal decomposition / combustion, in air, of solid-solution precursors of the types (M = Mg, Mn, Fe, Ni or Zn), $M_{1/3}Fe_{2/3}(N_2H_3COO)_2(N_2H_4)_2$ [203] (M = Mn, Co, Ni, Zn or Cd) and $M_{1/3}Co_{2/3}(N_2H_3COO)_2(N_2H_4)_2$ [204] (M = Mn, Fe, Ni, Zn or Cd), $M_{1/3}Co_{2/3}(C_6H_5CH_2COO)_2(N_2H_4)_2$ (M = Ni or Cd) [205]. Recently $Co_{0.8}Zn_{0.2}Fe_2O_4$ nanoparticles have been prepared by lower temperature thermal decomposition of $Co_{0.8}Zn_{0.2}Fe_2(C_4H_2O_4)_3.6N_2H_4$ [206].

The thermal decomposition of metal sulphite hydrazine hydrates [207], $Mg(HSO_3)_2N_2H_4.H_2O$ [208] and $MFe_2 (SO_3)_3 (N_2H_4)_6.2H_2O$ (M = Mg, Mn, Co, Ni or Zn) [209] have been reported by Budkuley and Patil. The decomposition of mixed metal sulphite hydrazines occurs at a lower temperature due to high exothermicity of the hydrazine decomposition in the complexes. Iron is also known to catalyse the decomposition of hydrazine. Hence these compounds undergo autocombustion (self sustained) once ignited. Sivasankar and Govindarajan [210] have reported the thermal decomposition behaviour of metal hydrazine sulphinates for the first time.

1.6. Objectives of the work

The present work has been taken up with the following objectives.

- To synthesise the complexes [M(cin)$_2$(N$_2$H$_4$)$_2$] (M= Ni, Co, Zn or Cd) from the corresponding nitrate hexahydrates, hydrazine hydrate and cinnamic acid.
- To prepare hetero bimetallic complexes [MFe$_2$(cin)$_3$(N$_2$H$_4$)$_3$] (M= Ni, Co, Zn or Cd) from the corresponding aqueous solutions of metal nitrate hexahydrates, ferrous sulphate, cinnamic acid and hydrazine hydrate.
- To synthesise hetero trimetallic complexes using aqueous solutions of metal nitrate hexahydrates, ferrous sulphate, cinnamic acid and hydrazine hydrate.
- To characterize the prepared complexes by analytical data, IR spectroscopy and TG-DTA/ TG-DSC.
- To analyze the surface morphology of the thermal decomposition product of the prepared complexes by using Scanning Electron Microscope (SEM).
- To study the compositional analysis of the final residue of thermally decomposed complexes by Energy Dispersive Spectrum (EDX).

1.7. Organization of study

The study is divided into six chapters.

Chapter I

This chapter deals with a general introduction of coordination complexes is given. Hydrazine chemistry is also discussed in detail. A brief review of literature related to the preparation methodology of metal hydrazine carboxylates is also presented. Objectives of the work are also presented.

Chapter II

The details of the chemicals and experimental techniques used in the synthesis and characterization of the samples under study are discussed in this chapter.

Chapter III

This chapter illustrates the synthesis of the complexes [M(cin)$_2$(N$_2$H$_4$)$_2$], (M= Ni, Co, Zn or Cd) and their characterisation.

Chapter IV

In this chapter, the synthesis of hetero bimetallic complexes and their characterization have been reported.

Chapter V

This chapter emphasises the synthesis and characterisation of hetero trimetallic complexes.

Chapter VI

This chapter details the summary and conclusions drawn from the present work. Further scope of the work is also discussed.

CHAPTER II
EXPERIMENTAL DETAILS

A brief description of the analytical procedures and physico-chemical techniques employed in the present investigations are presented in this chapter.

2.1. Materials

Analar grade chemicals were used as such without any further purification. Double distilled water was used in all of the studies. The following is the list of chemicals used in our present work.

1. 8-hydroxyquinaldine (2-methyloxine)
2. Absolute alcohol
3. Acetic acid
4. Ammonium chloride
5. Ammonium thiocyanate
6. Cadmium nitrate tetrahydrate
7. Carbon tetrachloride
8. Cinnamic acid – *trans*
9. Cobalt nitrate hexahydrate
10. Concentrated Ammonia
11. Diammonium hydrogenphosphate
12. Diethyl ether
13. Dimethyl Glyoxime
14. Ferrous sulphate
15. Hydrazine hydrate
16. Hydrochloric acid
17. Mercuric chloride
18. Mercury(II) chloride
19. Nickel nitrate hexahydrate
20. Nitric acid
21. N-Phenyl anthranilic acid

22. Potassium dichromate
23. Potassium iodate
24. Quinaldic acid
25. Stannous chloride
26. Sulphuric acid
27. Zinc nitrate hexahydrate

2.2. Analytical methods

2.2.1. Estimation of hydrazine in the complexes

The hydrazine content of the complexes was determined volumetrically using a standard KIO_3 (0.025 M) solution under Andrew's conditions [211].

$$IO_3^- + N_2H_4 + 2H^+Cl^- \longrightarrow ICl + N_2 + 3H_2O$$

1 ml of 0.025 M KIO_3 = 0.0008013 g of hydrazine.

In an iodimetry flask, 100 mg of the sample was dissolved in 10 mL of concentrated hydrochloric acid and 20 mL of distilled water. 5 mL of carbon tetrachloride was added to it. It was titrated against standard potassium iodate (0.025 M) solution from the burette. The solution was shaken well after the addition of each mL of KIO_3 solution. The end point was the disappearance of purple colour in the organic layer.

2.2.2 Estimation of metal ions in hetero bimetallic hydrazine cinnamates

Estimation of iron

The complex was thermally decomposed to its corresponding oxide, which was dissolved in conc. HCl. To this, ammonium chloride and conc. NH_3 were added. Reddish brown precipitate of ferric hydroxide formed was filtered by using a Whatmann filter paper, by washing with hot water.

The collected ferric hydroxide was dissolved in 5 mL of conc. HCl and boiled. To this hot solution, fresh stannous chloride was added in drops with stirring, until the yellow colour of the solution disappeared. Two or more drops of stannous chloride were added again and the solution was then cooled. Then about 5 mL of cold saturated solution (10%) of mercuric chloride was added rapidly with constant stirring. This was added to remove excess stannous chloride. A silky white precipitate of mercurous chloride formed was filtered off. To the turbid solution containing ferrous salt, about 20 mL of 2N H_2SO_4

and 15 drops of N-Phenyl anthranilic acid as indicator were added and the solution was titrated against standard potassium dichromate until there was a colour change from green to violet red.

Estimation of cobalt

After the separation of iron, the filtrate obtained was dissolved in 1 mole of mercury (II) chloride and 5 moles of ammonium thiocyanate in water. The blue salt of cobalt tetrathiocyanatomercurate (II) $Co[Hg(SCN)_4]$ formed was filtered, dried in a sintered crucible and then the cobalt content was estimated.

Estimation of nickel

Nickel was precipitated by the addition of an ethanolic solution of dimethylglyoxime and then a slight excess of aqueous ammonia solution. The precipitate obtained as nickel dimethylglyoximate was washed with cold water, dried and then nickel content was estimated.

Estimation of zinc

Zinc was precipitated as $Zn(C_{10}H_8ON)_2$ by the addition of 8-hydroxyquinaldine (2-methyloxine) in acetic acid solution. $Zn(C_{10}H_8ON)_2$ was dried and the content of zinc was estimated.

Estimation of cadmium

Quinaldic acid was used to precipitate cadmium as quinaldate. The precipitate was collected on a sintered-glass crucible, dried and zinc content was estimated.

2.2.3. Estimation of metals in hetero trimetallic hydrazine cinnamates

Ferric ion was removed and estimated for iron as illustrated in the procedure given in the previous section.

Co, Zn and Fe

Cobalt ion was removed as cobalt tetrathiocyanatomercurate(II) and estimated for cobalt. The interfering zinc ion was suppressed by the addition of phosphate. The filtrate obtained was estimated for zinc as $Zn(C_{10}H_8ON)_2$.

Ni, Zn and Fe

Nickel ion was removed as nickel dimethylglyoximate and estimated for nickel. The filtrate obtained was estimated for zinc as $Zn(C_{10}H_8ON)_2$.

2.3. Physico-chemical techniques

The instrumental techniques used in the present study are, Infrared spectroscopy (IR), Thermo Gravimetry-Differential Thermal Analysis/Differential Scanning Calorimetry (TG-DTA/TG-DSC), Scanning Electron Microscopy (SEM) and Energy Dispersive X-Ray Analysis (EDX).

2.3.1 Infrared spectral analysis

The infrared spectrum of a molecule is considered to be its unique and characteristic physical property. The basic interpretation of an IR spectrum leads to the characterization and identification of the sample. The IR spectrum is formed as a consequence of the absorption of electromagnetic radiation at frequencies that correlate to the vibration of specific sets of chemical bonds from within a molecule. Bonds with different bond lengths, strength, bending and torsional characteristics absorb different wavelengths of IR radiation, and the absorption maxima maybe characteristics of the types of linkages present; the absorption generally occurs in the 4000-400 cm^{-1} region of the spectrum.

Infrared radiation is capable of affecting both rotational and vibrational energy levels in the molecules [212, 213]. Some molecular vibrations are characteristic of the entire molecule, whereas others are associated with certain functional groups. IR spectra of solids are usually complex with a considerable number of peaks, each corresponding to a particular vibrational transition. Since a complete assignment of all the peaks to the specific vibrational modes is possible, straight forward identification of specific functional groups, covalently bonded linkage such as hydroxyl groups, trapped water, oxyanions, carbonates, nitrates, sulphates etc., and the modes of binding of the ligand to the metal ion can be made. As such, infrared spectroscopy has become an important technique in the elucidation of the structure of organic compounds and metal complexes. Plots of IR spectra show the frequency or wave number of incident radiation on the x-axis and percentage of transmittance on the y-axis. The wave number unit is used more often since it is directly proportional to the energy of vibration and such modern IR instruments are linear in cm^{-1} scale.

In the present study, the IR spectra of all the complexes were recorded by the KBr disc techniques using Perkin Elmer 597/1650 spectrophotometer.

2.3.2. Thermogravimetric analysis (TGA)

The thermal analysis is a technique in which the physical and chemical transitions of a substance are recorded as a function of temperature. In thermogravimetric analysis (TGA), the mass of the sample is recorded continuously as a function of temperature. TGA curves are specific for a given compound or system due to the unique sequence of physiochemical reactions that occur at definite temperatures and at rates controlled by the molecular structure. Changes in the mass of the sample take place as a result of the rupture and formation of various chemical bonds at higher temperatures, which leads to the evolution of volatile products or formation of reaction products.

In present study, the simultaneous TG-DTA experiments were carried out in Shimadzu DT40, Stanton 781 and STA 1500 thermal analyzers. Thermal analyses were carried out in air at the heating rate of 10°C per minute using 5-10 mg of the samples. Platinum cups were used as sample holders and alumina as reference. The temperature range was ambient to 700°C.

TG-DSC was performed using the Universal V4.5A TA Instrument in a temperature range of 50-700 °C with a heating rate of 20 °C/min.

2.3.3. Scanning Electron Microscopy (SEM) – Energy Dispersive X-ray analysis (EDX)

SEM is another powerful technique used to study the morphology (surface structures) of nanoparticles. SEM is able to provide 3-dimensional images of the objects, since it does not record the electrons passing thorough the specimen; on the other hand, it records the secondary electrons that are released from the sample as a result of interaction with falling electron beam. In SEM, high-energy electrons impinging on a solid sample experience elastic scattering by atomic nuclei and inelastic scattering by sample electrons. Inelastic scattering results in the transferring of energy from the high-energy electrons to the sample electrons. Thus, sample electrons can be excited to high-energy states. Some of the excited electrons travelling to the sample surface can be emitted out of the sample as secondary or Auger electrons. By collecting these low energy electrons, high-resolution surface images can be obtained to give morphological information about the sample.

Energy dispersive X-ray analysis is a chemical microanalysis technique used for elemental composition analysis. This technique is used in conjunction with scanning electron microscopy (SEM). The EDX technique detects X-rays emitted from the sample during bombardment by an electron beam to characterize the elemental composition. When the sample is bombarded by the SEM's electron beam, electrons are ejected from the atoms comprising the sample's surface. The resulting electron vacancies are filled by electrons from a higher state, and an X-ray is emitted to balance the energy difference between the two electronic states. The X-ray energy is characteristic of the element from which it was emitted. The spectrum of X-ray energy versus counts is evaluated to determine the elemental composition of the samples.

In the present study, SEM and EDX of the samples were performed with a HITACHI Model S-3000H.

CHAPTER III
SYNTHESIS AND CHARACTERIZATION OF METAL HYDRAZINE CINNAMATES [M(cin)$_2$(N$_2$H$_4$)$_2$] (M= Ni, Co, Zn or Cd)

Abstract

This chapter illustrates the synthesis of metal hydrazine cinnamates of the general formula [M(cin)$_2$(N$_2$H$_4$)$_2$] (M = Ni, Co, Zn or Cd) and its characterization by analytical data, IR spectroscopy and ThermoGravimetry - Differential Thermal Analysis (TG-DTA). The bidentate bridging nature of the hydrazine ligand and the monodentate coordination of the carboxylate groups to the metal ions in the complexes are known form the IR spectral data. The thermal decomposition of the prepared complexes has led to the formation of the corresponding metal oxides as confirmed by TG-DTA and EDX.

3.1. Synthesis and characterization of the complex [Co(cin)$_2$(N$_2$H$_4$)$_2$]

3.1.1. Synthesis of [Co(cin)$_2$(N$_2$H$_4$)$_2$]

[Co(cin)$_2$(N$_2$H$_4$)$_2$] was prepared by the addition of an aqueous solution (50 mL) of hydrazine hydrate (0.5 mL, 0.001 mol) and cinnamic acid (0.74 g 0.005 mol) to the corresponding aqueous solution (50 mL) of cobalt nitrate hexahydrate (0.73 g, 0.002 mol). A pink coloured precipitate was formed in a few minutes. The obtained precipitate was kept aside for an hour, then filtered and washed with water, alcohol followed by diethylether to remove adsorbed impurities and then dried at room temperature.

Co(NO$_3$)$_{2(aq)}$ + 2N$_2$H$_{4(aq)}$ + 2C$_6$H$_5$-CH=CH-COOH →

Co(C$_6$H$_5$-CH=CH-COO)$_2$(N$_2$H$_4$)$_2$ + 2HNO$_3$

3.1.2. Characterization of [Co(cin)$_2$(N$_2$H$_4$)$_2$]
3.1.2.1. Analytical data

The chemical formula [Co(cin)$_2$(N$_2$H$_4$)$_2$] has been assigned to the complex based on the observed percentage of hydrazine (14.90) and cobalt (14.00) which are found to match closely with the calculated values (15.35) and (14.13) for hydrazine and cobalt respectively.

3.1.2.2. IR spectral analysis

From the IR spectrum of the complex (Fig. 3.1), it is observed that the N-N stretching frequency seen at 962 cm^{-1} unambiguously proving the bidentate bridging nature of the hydrazine moieties [29]. The asymmetric and symmetric stretching frequencies of the carboxylate ions are seen at 1600 and 1400 cm^{-1}, respectively with the $\Delta\upsilon$ (υ_{asymm}- υ_{sym}) separation of 200 cm^{-1}, which indicate the monodentate linkage of the carboxylate groups. The band appearing at 3310 cm^{-1} can be attributed to the N-H stretching vibration.

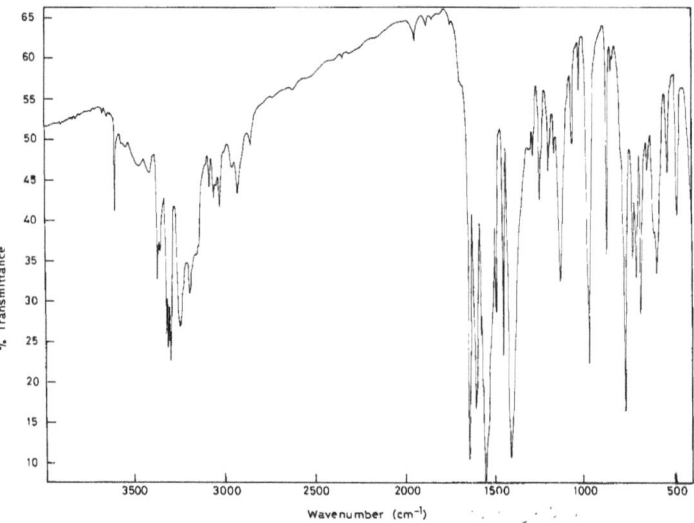

Fig. 3.1 IR spectrum of [Co(cin)$_2$(N$_2$H$_4$)$_2$]

3.1.2.3. Thermal analysis of [Co(cin)$_2$(N$_2$H$_4$)$_2$]

As can be observed from the simultaneous TG-DTA curves in Fig. 3.2, the complex loses weight in two particular steps. The first step is the attributed to dehydrazination of the two hydrazine molecules between 158-256 °C. The observed mass loss of 15.50% goes well with the calculated one which is 15.35%. The corresponding peak in DTA is observed as an exotherm within the temperature range of TG inflexion. The intermediate formed would be the corresponding metal carboxylate, which is not thermally stable. The major weight loss [60.00% (observed) and 60.45% (calculated)] on the TG curve from 251 to 470 °C is attributed to the second step involving the decarboxylation of the dehydrazinated complex, which gives Co_3O_4 as the final residue.

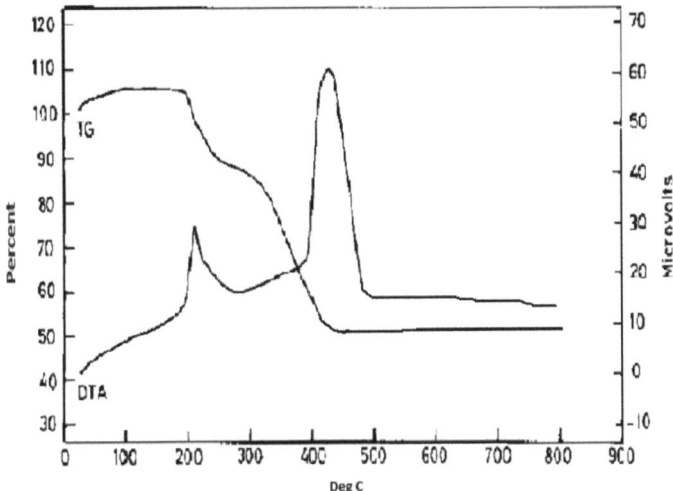

Fig. 3.2 TG-DTA curve of $[Co(cin)_2(N_2H_4)_2]$

3.1.2.4. SEM and EDX analysis

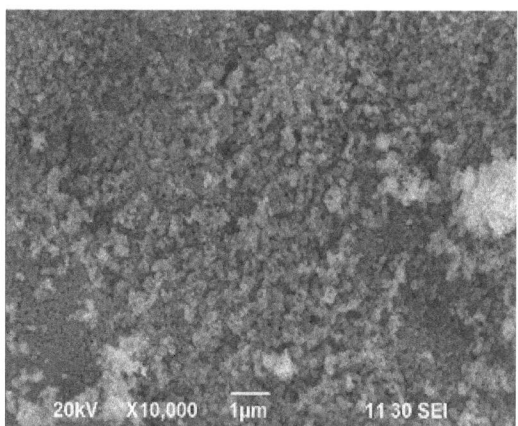

Fig. 3.3 SEM image of Co_3O_4

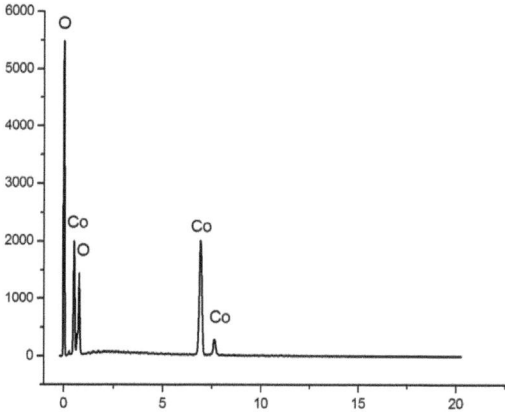

Fig. 3.4 EDX spectrum of Co_3O_4

Figures 3.3 and 3.4 show respectively the SEM image and the EDX spectrum of the thermal decomposition product of $[Co(cin)_2(N_2H_4)_2]$ viz., Co_3O_4. As can be seen from the EDX spectrum, the final product contains only the expected elements and no unwanted ones are seen.

3.2. Synthesis and characterization of the complex $[Ni(cin)_2(N_2H_4)_2]$

3.2.1. Synthesis of $[Ni(cin)_2(N_2H_4)_2]$

$[Ni(cin)_2(N_2H_4)_2]$ was prepared by the addition of an aqueous solution (50 ml) of hydrazine hydrate (0.5 ml, 0.001mol) and cinnamic acid (0.74 g, 0.005 mol) to the corresponding aqueous solution (50 ml) of nickel nitrate hexahydrate (0.73 g, 0.002 mol). The blue precipitate formed was kept undisturbed for an hour, then filtered and washed with water, alcohol followed by diethylether and air dried.

$Ni(NO_3)_{2(aq)} + 2N_2H_{4(aq)} + 2C_6H_5\text{-}CH\text{=}CH\text{-}COOH \rightarrow$

$Ni(C_6H_5\text{-}CH\text{=}CH\text{-}COO)_2(N_2H_4)_2 + 2HNO_3$

3.2.2. Characterization of $[Ni(cin)_2(N_2H_4)_2]$

3.2.2.1 Analytical data

A tentative chemical formula $[Ni(cin)_2(N_2H_4)_2]$ has been fixed to the complex, on the basis of the found values of hydrazine (15.00) and nickel (14.10) percentage which are found to best fit respectively with the calculated percentage values (15.36) and (14.08) for hydrazine and nickel.

3.2.2.2 IR spectral analysis

In the IR spectrum of $[Ni(cin)_2(N_2H_4)_2]$ shown in Fig. 3.5, a sharp band is noticed at 972 cm^{-1}, which corresponds to the N-N stretching frequency, explicitly proving that the hydrazine moieties are coordinated with the metal in a bidentate bridging fashion. The carboxylate ion asymmetric and symmetric stretching frequencies are observed at 1612 and 1384 cm^{-1}, respectively with the $\Delta\upsilon$ value of 288 cm^{-1}, indicating the monodentate liganding of –COOH groups. An intense band at 3305 cm^{-1} shows the N-H stretching in the prepared complex.

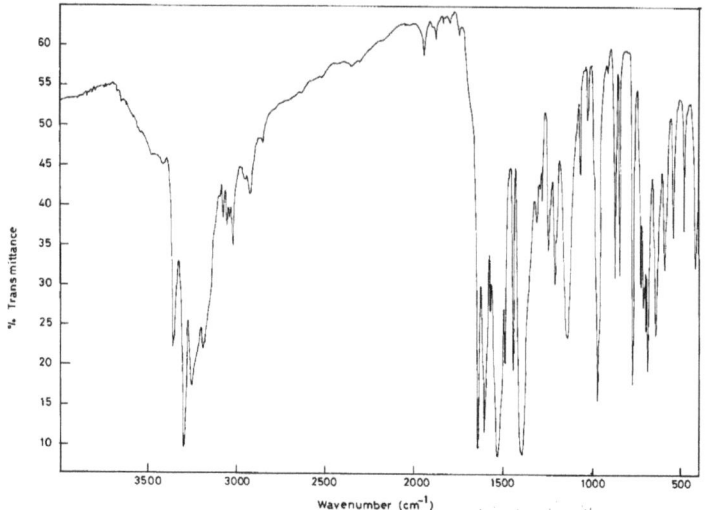

Fig. 3.5 IR spectrum of [Ni(cin)$_2$(N$_2$H$_4$)$_2$]

3.2.2.3 Thermal analysis

Fig. 3.6 shows the TG-DTA curves of the complex. The weight loss occurs in two steps, the initial one being the exothermic dehydrazination of the complex between 158 °C and 256 °C, which results in the formation of an unstable metal carboxylate intermediate. The final step which takes place from 251 to 470 °C is ascribed to the decarboxylation of the dehydrazinated complex, giving metallic Ni as the final residue.

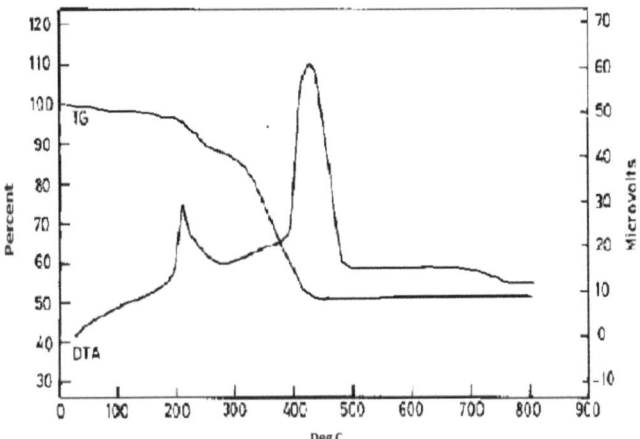

Fig. 3.6 TG-DTA curve of [Ni(cin)$_2$(N$_2$H$_4$)$_2$]

3.2.2.4. SEM and EDX analysis

Fig. 3.7 SEM image of metallic Ni

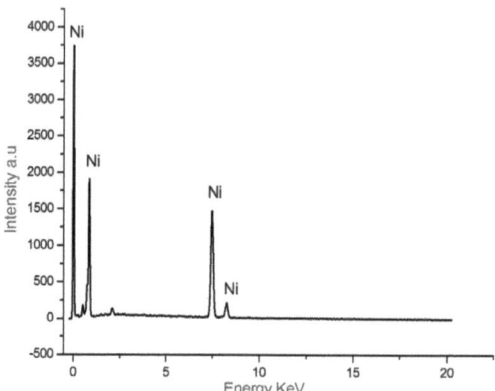

Fig. 3.8 EDX spectrum of metallic Ni

Figures 3.7 and 3.8 show respectively the SEM image and the EDX spectrum of the thermal decomposition product of [Ni(cin)$_2$(N$_2$H$_4$)$_2$] viz., Ni metal. As can be seen from the EDX spectrum, the final product contains only Ni.

3.3 Synthesis and characterization of [Zn(cin)$_2$(N$_2$H$_4$)$_2$]

3.3.1. Synthesis of [Zn(cin)$_2$(N$_2$H$_4$)$_2$]

This was prepared by the addition of an aqueous solution (50 mL) of hydrazine hydrate (2 mL, 0.4 mol) and cinnamic acid (0.74g 0.005 mol) to the corresponding aqueous solution (50 mL) of zinc nitrate hexahydrate (0.7437 g, 0.002 mol). The white precipitate formed was kept aside for an hour, then filtered and washed with water, alcohol followed by diethylether. The washing procedure was repeated two more times and then dried at room temperature.

Zn(NO$_3$)$_{2(aq)}$ + 2N$_2$H$_{4(aq)}$ + 2C$_6$H$_5$-CH=CH-COOH →

$$\text{Zn(C}_6\text{H}_5\text{-CH=CH-COO)}_2\text{(N}_2\text{H}_4\text{)}_2 + 2\text{HNO}_3$$

3.3.2. Characterization of [Zn(cin)$_2$(N$_2$H$_4$)$_2$]

3.3.2.1. Analytical data

The chemical formula [Zn(cin)$_2$(N$_2$H$_4$)$_2$] has been assigned to the prepared complex, based on the observed and calculated percentage values of hydrazine and zinc, which are found to match closely with the calculated values (Table 1).

Compound	Hydrazine (%)		Zinc (%)	
Zn(cin)$_2$(N$_2$H$_4$)$_2$	Found	Calculated	Found	Calculated
	14.50	15.12	15.00	15.44

Table 3.1- Analytical data

3.3.2.2. IR spectral analysis

From the IR spectrum of the complex (Fig. 3.9), it is observed that the N-N stretching frequency is seen at 968 cm^{-1}, which clearly proves the bidentate bridging nature of the hydrazine moieties. The asymmetric and symmetric stretching frequencies of the carboxylate ions are seen at 1600 and 1406 cm^{-1}, respectively with the $\Delta\upsilon$ (υ_{asymm}- υ_{sym}) separation of 194 cm^{-1}, which indicate the monodentate linkage of both the carboxylate groups. The N-H stretching is observed at 3320 cm^{-1}.

3.3.2.3. Thermal analysis

As can be observed from Fig. 3.10, the complex loses weight in three particular steps. The first step is the loss of one of the hydrazine molecules from 142-201 °C. The corresponding peak in DTA is observed as an endotherm at 173 °C. The second step, being the loss of another hydrazine molecule, is also endothermic and observed at 219 °C in DTA. The major weight loss of 80% on the TG curve from 231 to 525 °C is attributed to the third step involving the decarboxylation of the dehydrazinated complex, which gives zinc oxide as the final residue.

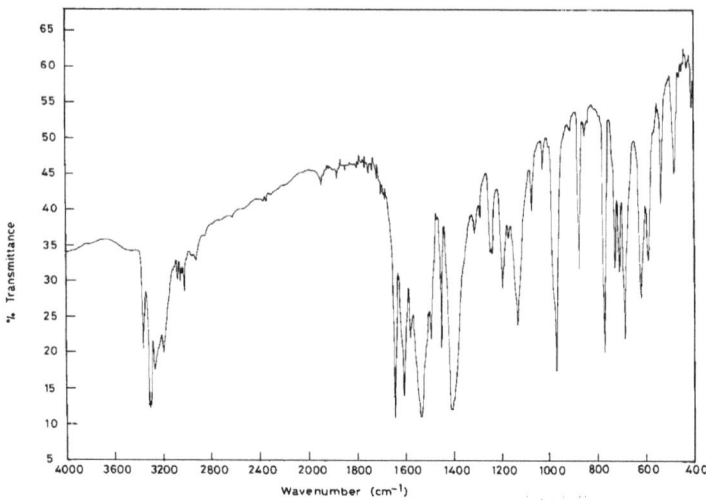

Fig. 3.9 IR spectrum of [Zn(cin)$_2$(N$_2$H$_4$)$_2$]

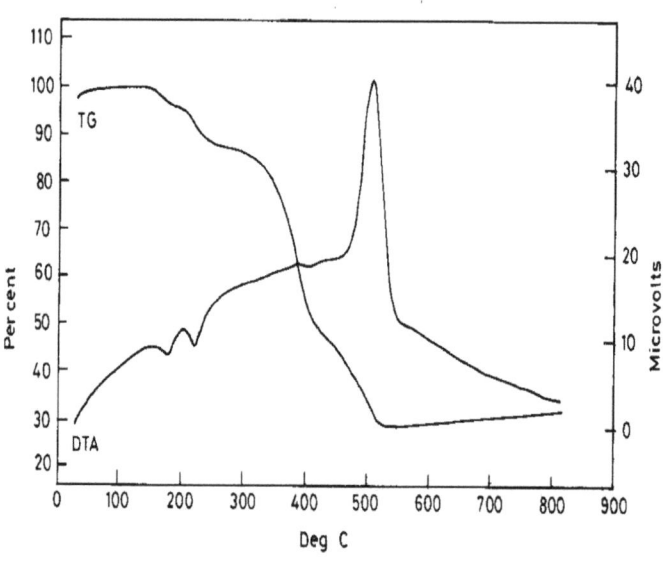

Fig. 3.10 TG-DTA curve of [Zn(cin)$_2$(N$_2$H$_4$)$_2$]

3.3.2.4. SEM and EDX analysis

Figures 3.11 and 3.12 show respectively the SEM image and the EDX spectrum of the thermal decomposition product of $[Zn(cin)_2(N_2H_4)_2]$ viz., ZnO. The EDX spectrum clearly pictures the actual composition of the final product. No other impurity peaks are detected.

Fig. 3.11 SEM image of ZnO

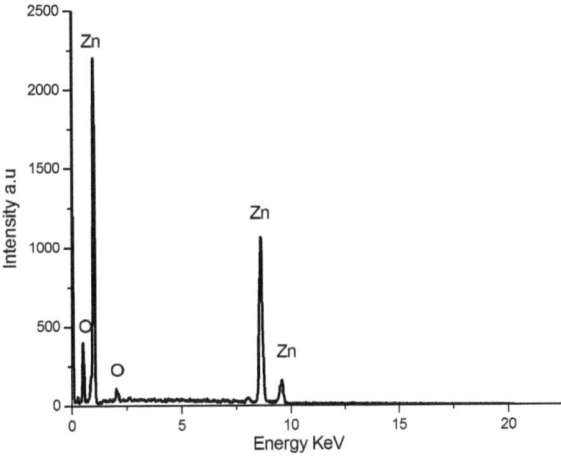

Fig. 3.12 EDX spectrum of ZnO

3.4. Synthesis and characterization of [Cd(cin)$_2$(N$_2$H$_4$)$_2$]

3.4.1. Synthesis of [Cd(cin)$_2$(N$_2$H$_4$)$_2$]

This complex was prepared by the addition of an aqueous solution (50 mL) of hydrazine hydrate (1 mL, 0.01 mol) and cinnamic acid (0.74g 0.055 mol) to the corresponding aqueous solution (50 mL) of cadmium nitrate hexahydrate (0.77 g, 0.002 mol) . The white coloured precipitate formed in a few minutes was filtered after an hour. Washings were done with water and alcohol followed by diethylether. The sample was then air dried and stored.

$$Cd(NO_3)_{2(aq)} + 2N_2H_{4(aq)} + 2C_6H_5\text{-CH=CH-COOH} \rightarrow$$
$$Cd(C_6H_5\text{-CH=CH-COO})_2(N_2H_4)_2 + 2HNO_3$$

3.4.2. Characterization of [Cd(cin)$_2$(N$_2$H$_4$)$_2$]

3.4.2.1. Analytical data

The chemical formula [Cd(cin)$_2$(N$_2$H$_4$)$_2$] has been assigned to the complex, on the basis of the observed percentage of hydrazine (13.70) and cadmium (23.60) which are found to match closely with the calculated values (13.61) and (23.90) for hydrazine and cadmium respectively.

3.4.2.2. IR spectral analysis

The IR spectrum of the [Cd(cin)$_2$(N$_2$H$_4$)$_2$] (Fig. 3.13) displays a band at 962 cm^{-1}, characteristic of N-N stretching frequency. This clearly proves the bidentate bridging nature of hydrazine. The asymmetric and symmetric stretching frequencies of the carboxylate ions are seen at 1600 and 1396 cm^{-1}, respectively with the $\Delta \upsilon$ (υ_{asymm}- υ_{sym}) separation of 204 cm^{-1}, which indicate the monodentate linkage of the carboxylate groups. The band at 3300 cm^{-1} can be assigned to the N-H stretching.

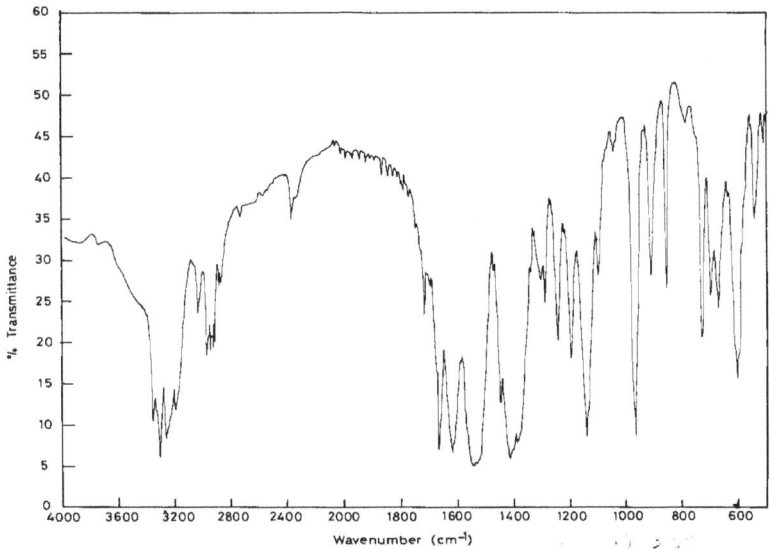

Fig. 3.13 IR spectrum of [Cd(cin)$_2$(N$_2$H$_4$)$_2$]

3.4.2.3. Thermal analysis

The simultaneous TG-DTA curves are shown in Fig. 3.14. The weight changes of the sample can be divided into three regions. The first step is the endothermic dehydrazination of the complex, which takes place between 166-297°C. In the second step, the unstable cadmium cinnamate gives cadmium acetate as the intermediate exothermically in the temperature range 297-395 °C. Our attempt to separate the cadmium acetate intermediate was unsuccessful due to its continuous decomposition. The final residue formed in the third step was analyzed as CdO which revealed thermal stability from 700 to 900 °C.

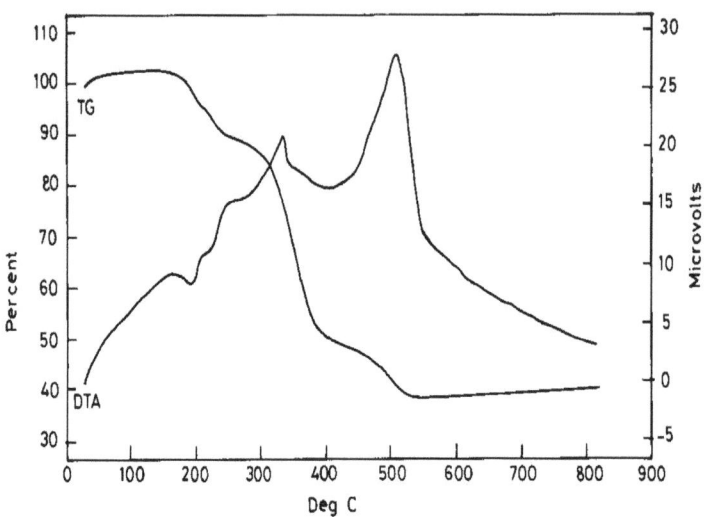

Fig. 3.14 TG-DTA curve of [Cd(cin)$_2$(N$_2$H$_4$)$_2$]

3.4.2.4. SEM and EDX analysis

The SEM picture in Fig. 3.15 clearly shows randomly distributed rock-candy like structures. The cohesion of particles is due to the magnetic attraction. EDX spectrum of CdO is presented in Fig. 3.16, which evinces that the product is mainly composed of CdO.

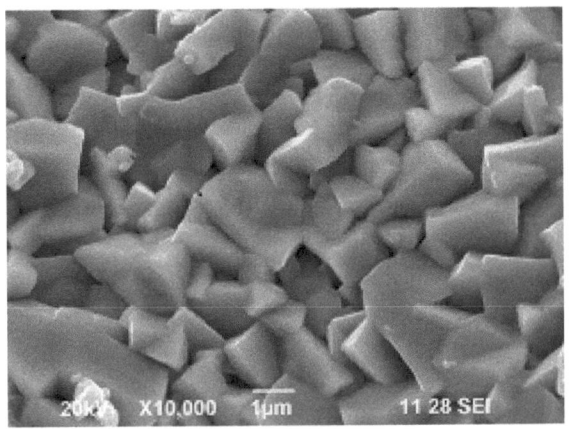

Fig. 3.15 SEM image of CdO

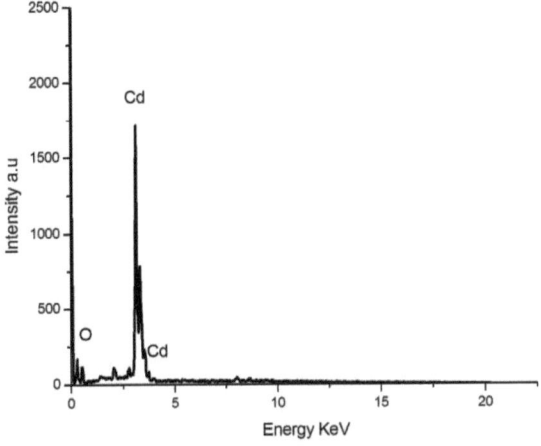

Fig. 3.16 EDX spectrum of CdO

3.5. Structure of the complexes [M(cin)$_2$(N$_2$H$_4$)$_2$] (M=Co, Ni, Zn or Cd)

On the basis of the analytical data, IR spectral study and thermal analysis, a tentative structure has been proposed for the prepared complexes. The proposed polymeric structure of [M(cin)$_2$(N$_2$H$_4$)$_2$] (M=Co, Ni, Zn or Cd) is given as follows, in which the hydrazine moieties are coordinated to the metal atom in a bidentate bridging fashion and the carboxylate groups in a unidentate manner.

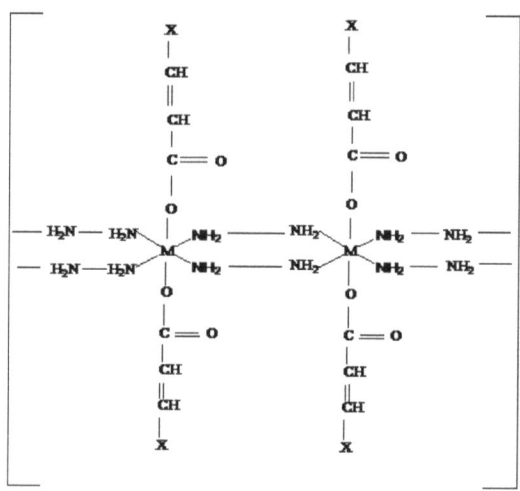

Conclusions

Transition metal complexes having the formula [M(cin)$_2$(N$_2$H$_4$)$_2$] (M=Co, Ni, Zn or Cd) were synthesized by a simple precipitation method using the corresponding metal nitrate hexahydrates, hydrazine and cinnamic acid. They were characterized by analytical data, IR spectra and TG-DTA. The IR spectra of all the complexes evinced the bidentate bridging and monodentate coordination of hydrazine and carboxylate groups respectively. Thermal decomposition of the complexes produced their respective oxides (Co$_3$O$_4$, ZnO and CdO) except [Ni(cin)$_2$(N$_2$H$_4$)$_2$], which yielded metallic Ni as the final residue, which was confirmed by TG-DTA and EDX. SEM images reported the morphology of the oxide samples.

CHAPTER IV
SYNTHESIS AND CHARACTERIZATION OF HETERO BIMETALLIC HYDRAZINE CINNAMATES [MFe2(cin)3(N2H4)3] (M= Ni, Co, Zn or Cd)

Abstract

The complexes having the general formula $[MFe_2(cin)_3(N_2H_4)_3]$ have been prepared form their corresponding metal salts, cinnamic acid and hydrazine hydrate. Their characterization has been done by analytical analysis, IR spectral study and TG-DSC. The IR spectral data shows the bidentate bridging nature of the hydrazine moieties and the unidentate linkage of the carboxylate groups. The TG-DSC patterns of all the complexes show multistep decomposition, yielding their respective mixed metal oxides as the final residue.

4.1. Synthesis and characterization of $[CoFe_2(cin)_3(N_2H_4)_3]$

4.1.1. Synthesis of $[CoFe_2(cin)_3(N_2H_4)_3]$

This was prepared by the addition of an aqueous solution (50 mL) of hydrazine hydrate (1 mL, 0.02 mol) and cinnamic acid (1.18 g, 0.0079 mol) to the corresponding aqueous solution (50 mL) of cobalt nitrate hexahydrate (0.58 g, 0.0018 mol) and ferrous sulphate heptahydrate (2.22 g, 0.0079 mol). The brown orange product formed in a few minutes was kept aside for an hour, then filtered and washed with water, alcohol followed by diethylether and air dried.

4.1.2. Characterization of $[CoFe_2(cin)_3(N_2H_4)_3]$

4.1.2.1. Analytical data

Based on the observed percentage of hydrazine (13.09), cobalt (24.32) and iron (47.17) which are found to match closely with the calculated values (13.64), (25.11) and (47.60) for hydrazine, cobalt and iron respectively, the chemical formula $[CoFe_2(cin)_3(N_2H_4)_3]$ has been tentatively assigned to the complex.

4.1.2.2. IR spectral analysis

In the IR spectrum of $[CoFe_2(cin)_3(N_2H_4)_3]$ (Fig.4.1), the N-N stretching frequency is seen at 972 cm^{-1}, explicitly proving the bidentate bridging nature of the hydrazine moieties [15]. The asymmetric and symmetric stretching frequencies of the carboxylate ions are seen at 1639 and 1411 cm^{-1}, respectively with the $\Delta\upsilon$ $(\upsilon_{asymm}-\upsilon_{sym})$ separation of 228 cm^{-1}, which confirm the monodentate linkage of the carboxylate groups. The N-H stretching is observed at 3367 cm^{-1}.

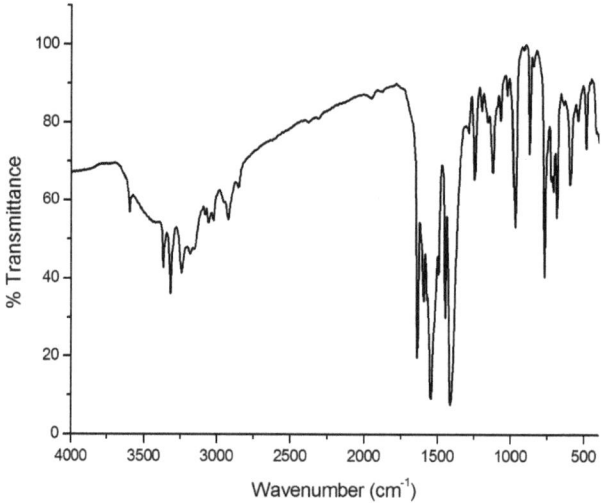

Fig. 4.1 IR spectrum of [CoFe$_2$(cin)$_3$(N$_2$H$_4$)$_3$]

4.1.2.3. Thermal analysis

As can be observed from Fig. 4.2, the complex loses weight in three particular steps. The first step is the removal of two of the hydrazine molecules between room temperature and 210 °C with a weight loss of 9%. The corresponding peak in DSC is observed as an exotherm. The second step is the loss of the remaining hydrazine molecule and one of the three carboxylate groups, which takes place between 210 and 400 °C. The weight loss from 400 to 610 °C is attributed to the third step involving the decarboxylation of the remaining two carboxylate moieties in the dehydrazinated complex, which gives CoFe$_2$O$_4$ as the final residue.

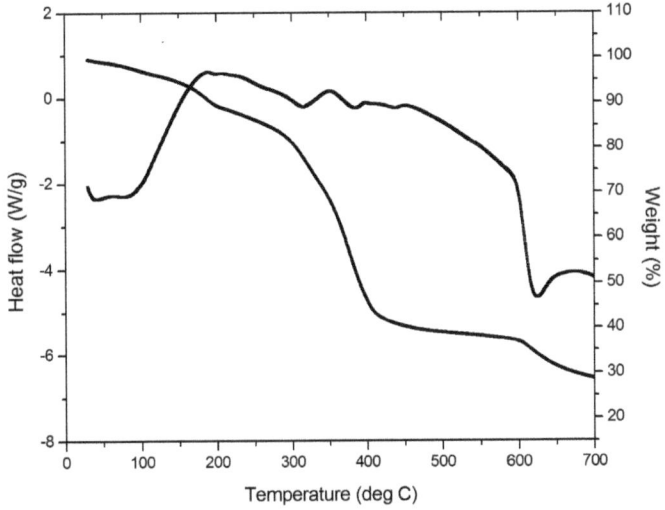

Fig.4.2 TG-DSC Curve of [CoFe$_2$(cin)$_3$(N$_2$H$_4$)$_3$]

4.1.2.4. SEM and EDX analysis

Fig.4.3 SEM image of CoFe$_2$O$_4$

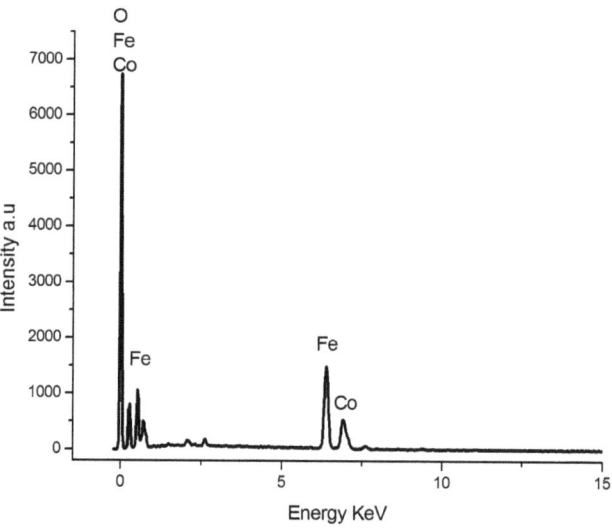

Fig. 4.4 EDX spectrum of CoFe$_2$O$_4$

The SEM picture in Fig. 4.3 clearly shows that the sample is mostly formed of nano-sized homogenous grains with the presence of some agglomerated particles. EDX spectrum is presented in Fig. 4.4, which furnishes the chemical compositional analysis of the thermal decomposition product CoFe$_2$O$_4$.

4.2. Synthesis and characterization of [NiFe$_2$(cin)$_3$(N$_2$H$_4$)$_3$]

4.2.1. Synthesis of [NiFe$_2$(cin)$_3$(N$_2$H$_4$)$_3$]

This was prepared by the addition of an aqueous solution (50 mL) of hydrazine hydrate (1 mL, 0.02 mol) and cinnamic acid (1.18 g, 0.0079 mol) to the corresponding aqueous solution (50 mL) of nickel nitrate hexahydrate (0.58 g, 0.0019 mol) and ferrous sulphate heptahydrate (2.22 g, 0.0079 mol). The brown orange product formed was left undisturbed for an hour and then filtered at the pump washed with water, alcohol followed by diethylether and dried in air.

4.2.2. Characterization of [NiFe$_2$(cin)$_3$(N$_2$H$_4$)$_3$]

4.2.2.1. Analytical data

Based on the observed percentage of hydrazine (13.05), nickel (24.79) and iron (46.99) which are found to match closely with the calculated values (13.65), (25.04) and (47.65) for hydrazine, nickel and iron respectively, the chemical formula NiFe$_2$(cin)$_3$(N$_2$H$_4$)$_3$ has been tentatively assigned to the prepared complex.

4.2.2.2. IR spectral analysis

The IR spectrum of [NiFe$_2$(cin)$_3$(N$_2$H$_4$)$_3$] is displayed in Fig.4.5, from which the N-N stretching frequency is observed at 972 cm^{-1}, which clearly proves that hydrazine is coordinated to the metals in a bidentate bridging nature. υ_{asymm} and υ_{sym} of the carboxylate ions are seen respectively at 1639 and 1404 cm^{-1}, with the $\Delta\upsilon$ value of 235 cm^{-1}, which show the monodentate linkage of the carboxylate groups. At 3356 cm^{-1}, a sharp band is noticed which is assigned to the N-H stretching frequency.

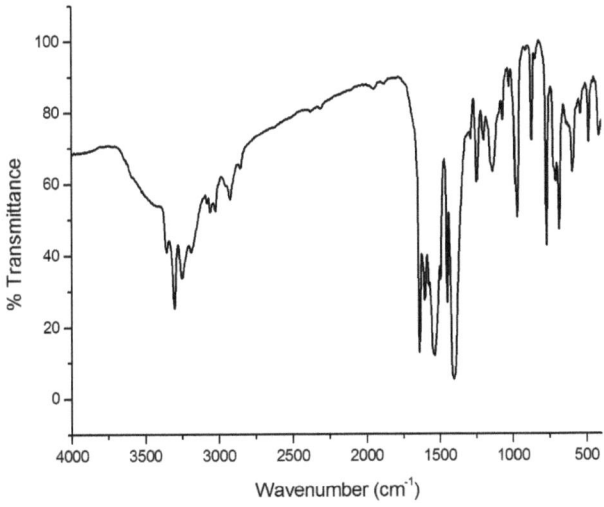

Fig. 4.5 IR spectrum of [NiFe$_2$(cin)$_3$(N$_2$H$_4$)$_3$]

4.2.2.3. Thermal analysis

As can be observed from Fig. 4.6, the complex loses weight in two particular steps. The first step is the dehydrazination of the compound between room temperature and 255 °C with a weight loss of 14%. The corresponding peak in DSC is observed as an endotherm. The major weight loss of 60% on the TG curve from 255 to 390 °C is attributed to the second step involving the decarboxylation of the dehydrazinated complex, which gives $NiFe_2O_4$ as the final residue.

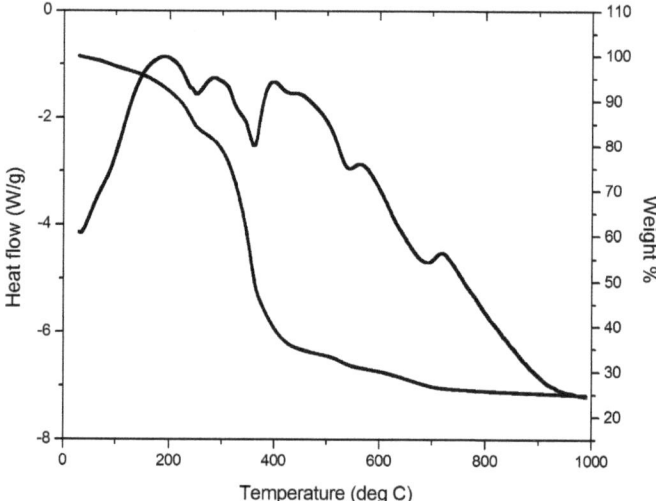

Fig.4.6 TG-DSC curve of $[NiFe_2(cin)_3(N_2H_4)_3]$

4.2.2.4. SEM and EDX analysis

Fig. 4.7 SEM micrograph of NiFe$_2$O$_4$

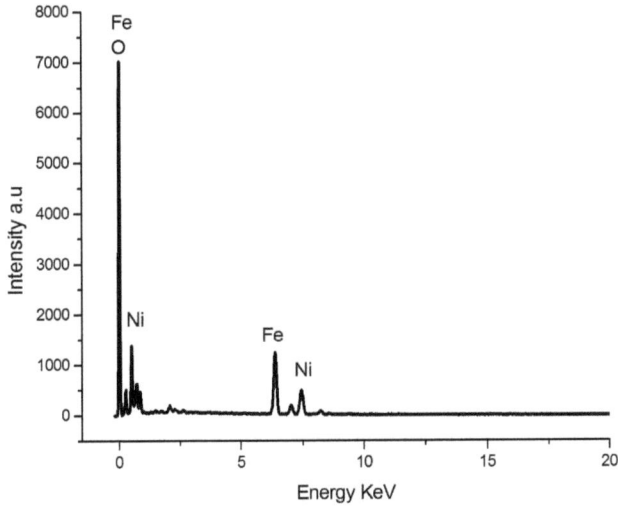

Fig. 4.8 EDX spectrum of NiFe$_2$O$_4$

The SEM picture in Fig. 4.7 evidently shows grains with smaller size and coalescence of particles. EDX spectrum of $NiFe_2O_4$ is presented in Fig.4.8 which provides the chemical composition of $NiFe_2O_4$.

4.3. Synthesis and characterization of $[ZnFe_2(cin)_3(N_2H_4)_3]$

4.3.1. Synthesis of $[ZnFe_2(cin)_3(N_2H_4)_3]$

This complex was prepared by adding an aqueous solution (50 mL) of hydrazine hydrate (1 mL, 0.02 mol) and cinnamic acid (1.18g, 0.0079 mol) to the aqueous solution (50 mL) containing zinc nitrate hexahydrate (0.058 g, 0.0019 mol) and ferrous sulphate heptahydrate (2.22g, 0.0079 mol). The brown orange product precipitated in a few minutes was allowed to stand for an hour and then filtered at the pump. Washings were done two or three times with water and alcohol followed by diethylether. The sample was then dried at room temperature and stored.

4.3.2. Characterization of $[ZnFe_2(cin)_3(N_2H_4)_3]$

4.3.2.1. Analytical data

Based on the observed percentage of hydrazine (13.38), zinc (26.84) and iron (45.99) which are found to match closely with the calculated values (13.52), (26.97) and (46.62) for hydrazine, zinc and iron respectively, the chemical formula $[ZnFe_2(cin)_3(N_2H_4)_3]$ has been tentatively assigned to the complex.

4.3.2.2. IR spectral analysis

The infrared spectrum of $[ZnFe_2(cin)_3(N_2H_4)_3]$ is given in Fig. 4.9. A band at 975 cm^{-1} corresponds to the N-N stretching frequency, from which the bidentate bridging nature of hydrazine to the metal atoms is clearly known. The asymmetric and symmetric stretching frequencies of the carboxylate ions at 1639 and 1411 cm^{-1} respectively with the Δυ separation of 228 cm^{-1}, indicate the monodentate linkage of the carboxylate groups. The N-H stretching is seen at 3371 cm^{-1}.

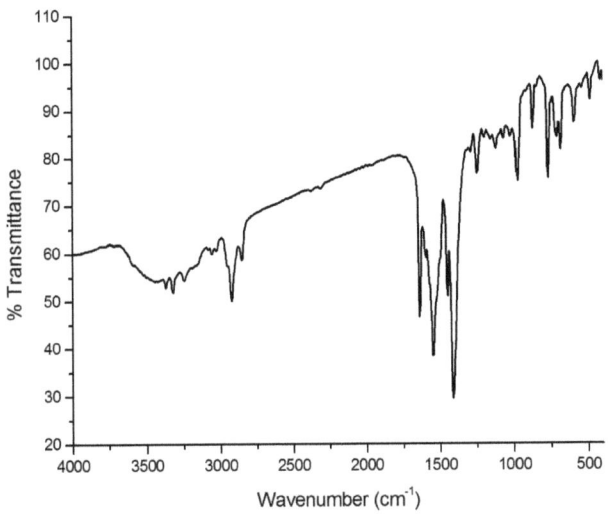

Fig. 4.9 IR spectrum of [ZnFe$_2$(cin)$_3$(N$_2$H$_4$)$_3$]

4.3.2.3. Thermal analysis

Fig.4.10 pictures the TG-DSC curve of the [ZnFe$_2$(cin)$_3$(N$_2$H$_4$)$_3$]. The weight loss takes place in two distinct steps. The first step is the removal of two of the molecules of hydrazine in the complex between room temperature and 175 °C with a weight loss of 9%. The corresponding peak in DSC is observed as an exotherm. The major weight loss of 60% on the TG curve from 175 to 430 °C is attributed to the second step involving the complete dehydrazination and decarboxylation of the complex, which gives ZnFe$_2$O$_4$ as the final residue.

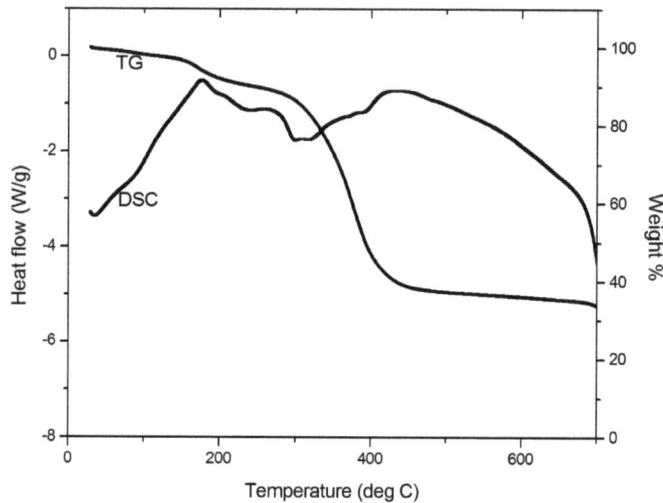

Fig.4.10 TG-DSC curve of [ZnFe$_2$(cin)$_3$(N$_2$H$_4$)$_3$]

4.3.2.4. SEM and EDX analysis

The SEM pictures in Fig. 4.11 clearly show randomly distributed grains with very smaller size and agglomeration of particles.

The composition of the sample is known from its EDX spectrum which is presented in Fig. 4.12.

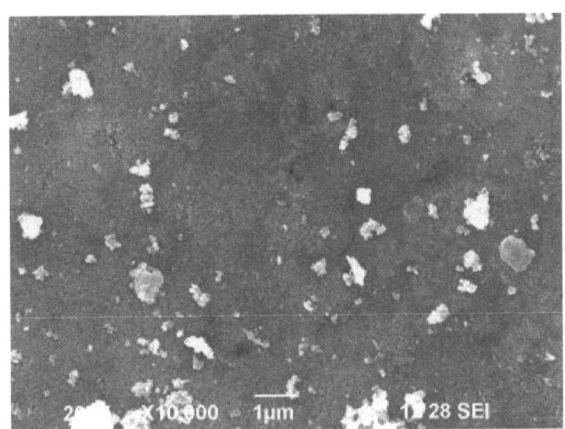

Fig. 4.11 SEM image of ZnFe$_2$O$_4$

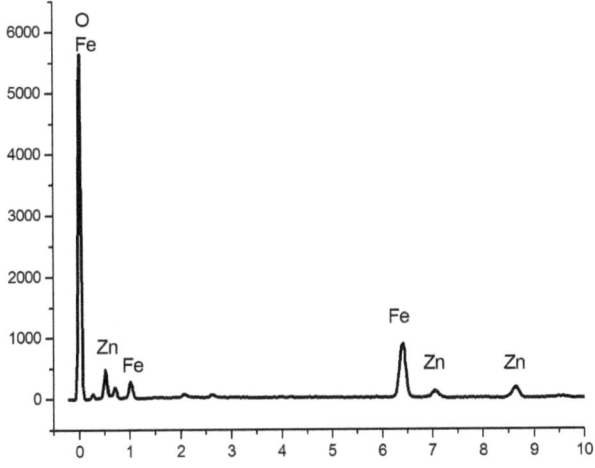

Fig. 4.12 EDX spectrum of ZnFe$_2$O$_4$

4.4. Synthesis and characterization of [CdFe$_2$(cin)$_3$(N$_2$H$_4$)$_3$]

4.4.1. Synthesis of [CdFe$_2$(cin)$_3$(N$_2$H$_4$)$_3$]

[CdFe$_2$(cin)$_3$(N$_2$H$_4$)$_3$] was prepared by the addition of an aqueous solution (50 mL) of hydrazine hydrate (1 mL, 0.02 mol) and cinnamic acid (1.18 g, 0.0079 mol) to the corresponding aqueous solution (50 mL) of cadmium nitrate hexahydrate (0.58 g, 0.0018 mol) and ferrous sulphate heptahydrate (2.22 g, 0.0079 mol). The brown orange product formed was kept aside for one hour, then filtered and washed with water, alcohol followed by diethylether and air dried.

4.4.2. Characterization of [CdFe$_2$(cin)$_3$(N$_2$H$_4$)$_3$]

4.4.2.1. Analytical data

Based on the observed percentage of hydrazine (12.09), cadmium (38.90) and iron (38.28) which are found to match closely with the calculated values (12.68), (39.01) and (38.77) for hydrazine, cadmium and iron respectively, the chemical formula [CdFe$_2$(cin)$_3$(N$_2$H$_4$)$_3$] has been given to the complex prepared.

4.4.2.2. IR spectral analysis

From the IR spectrum of [CdFe$_2$(cin)$_3$(N$_2$H$_4$)$_3$] (Fig. 4.13), it is observed that the N-N stretching frequency is seen at 972 cm^{-1}, which unambiguously proves the bidentate bridging nature of the hydrazine ligand. The asymmetric and symmetric stretching frequencies of the carboxylate ions are seen at 1639 and 1411 cm^{-1}, respectively with the $\Delta\upsilon$ (υ_{asymm}- υ_{sym}) separation of 228 cm^{-1}, which indicate the monodentate linkage of the carboxylate groups. The N-H stretching is observed at 3363 cm^{-1}.

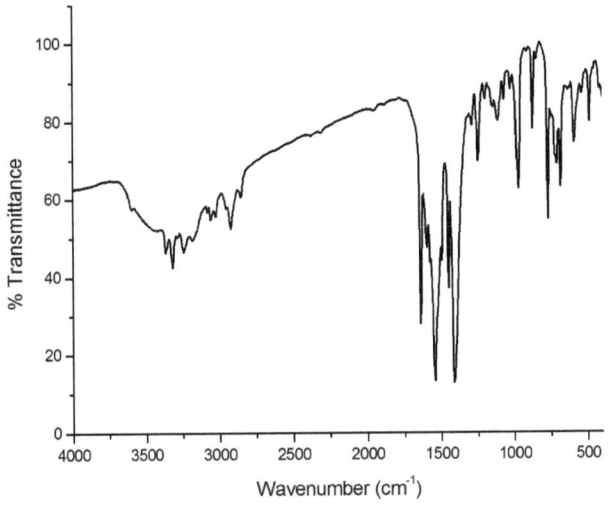

Fig. 4.13 IR spectrum of [CdFe$_2$(cin)$_3$(N$_2$H$_4$)$_3$]

4.4.2.3. Thermal analysis

As can be observed from Fig. 4.14, the [CdFe$_2$(cin)$_3$(N$_2$H$_4$)$_3$] loses weight in three particular steps. The first step is the removal of one of the hydrazine molecules between room temperature and 150 °C with a weight loss of 5%. The corresponding peak in DSC is observed as an endotherm. The second step is complete dehydrazination of the complex which takes place between 150 and 220 °C. The major weight loss of 60% on the TG curve from 220 to 410 °C is attributed to the third step involving the decarboxylation of the dehydrazinated complex, which gives CdFe$_2$O$_4$ as the final residue.

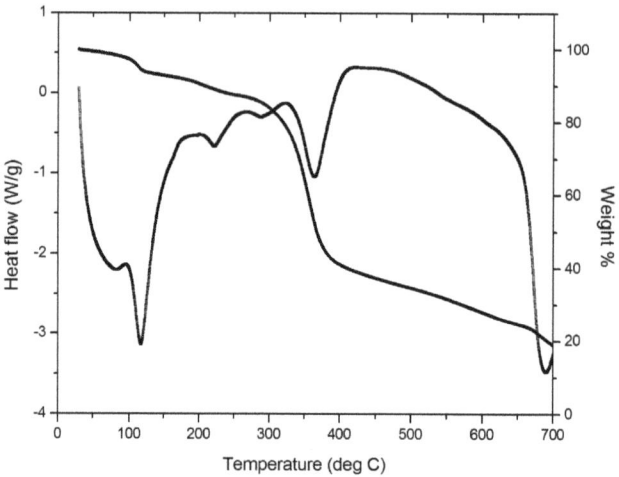

Fig.4.14 TG-DSC Curve of [CdFe$_2$(cin)$_3$(N$_2$H$_4$)$_3$]

4.4.2.4. SEM and EDX analysis

Fig.4.15 clearly shows randomly distributed grains with fairly uniform size and agglomeration of particles containing some voids. This observation could be attributed to the release of large amount of gases during decomposition. EDX spectrum of CdFe$_2$O$_4$ nanoparticles is presented in Fig.4.16.

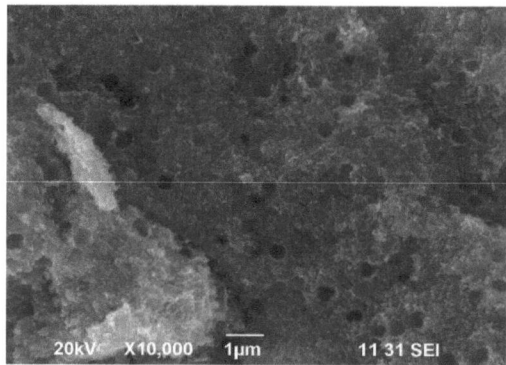

Fig.4.15 SEM image of CdFe$_2$O$_4$

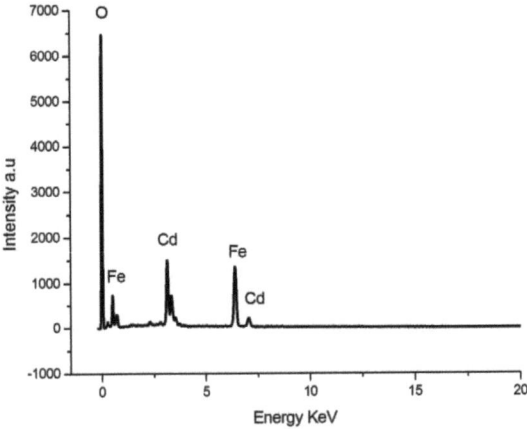

Fig. 4.16 EDX spectrum of CdFe$_2$O$_4$

Conclusions

The syntheses of the metal complexes with general formula [MFe$_2$(cin)$_3$(N$_2$H$_4$)$_3$] (M= Ni, Co, Zn or Cd) were achieved by a simple precipitation route. The prepared complexes were characterized by analytical data, IR spectra and TG-DSC. The complexes, on thermal decomposition, produced their respective oxides i.e. ferrites, which was confirmed by EDX. All the metal oxide samples possessed different morphologies, which were thoroughly studied by SEM.

CHAPTER V
SYNTHESIS AND CHARACTERIZATION OF HETERO TRIMETALLIC HYDRAZINE CINNAMATES

Abstract

Hetero trimetallic complexes have been prepared form their corresponding metal salts, cinnamic acid and hydrazine hydrate. Their characterization has been done by analytical analysis, IR spectral study and TG-DSC. The IR spectral data shows the bidentate bridging nature of the hydrazine moieties and the unidentate linkage of the carboxylate groups. The TG-DSC patterns of all the complexes show multistep decomposition, yielding their respective mixed metal oxides as the final residue.

5.1. Synthesis and characterization of [Ni$_{0.25}$Co$_{0.75}$Fe$_2$(cin)$_3$(N$_2$H$_4$)$_5$]

5.1.1. Synthesis of [Ni$_{0.25}$Co$_{0.75}$Fe$_2$(cin)$_3$(N$_2$H$_4$)$_5$]

This was prepared by the addition of an aqueous solution (50 mL) of hydrazine hydrate (1.6 mL, 0.03 mol) and cinnamic acid (1.18 g, 0.0079 mol) to the aqueous solution (50 mL) of cobalt nitrate hexahydrate (0.4656 g, 0.0015 mol), nickel nitrate hexahydrate (0.23 g, 0.0007 mol) and ferrous sulphate heptahydrate (1.11 g, 0.0039 mol). After complete reaction, a brown orange product was formed, which was filtered after an hour. Washings were done with water and alcohol followed by diethylether. The wet sample was dried in air and the dried sample was stored in a desiccator.

5.1.2. Characterization of [Ni$_{0.25}$Co$_{0.75}$Fe$_2$(cin)$_3$(N$_2$H$_4$)$_5$]

5.1.2.1. Analytical data

On the basis of the observed percentage of hydrazine (20.23), nickel (6.19), cobalt (19.09) and iron (46.97) which are found to match closely with the calculated values (20.84), (6.25), (18.84) and (47.61) for hydrazine, nickel, cobalt and iron respectively, the chemical formula [Ni$_{0.25}$Co$_{0.75}$Fe$_2$(cin)$_3$(N$_2$H$_4$)$_5$] has been fixed to the complex.

5.1.2.2. IR spectral analysis

The IR spectrum of [Ni$_{0.25}$Co$_{0.75}$Fe$_2$(cin)$_3$(N$_2$H$_4$)$_5$] (Fig. 5.1) displays an intensive band at 972 cm^{-1} due to the N-N stretching frequency, which unambiguously proves the bidentate bridging nature of the hydrazine molecules. The asymmetric and symmetric stretching frequencies of the carboxylate ions are seen at 1639 and 1411 cm^{-1}, respectively with the $\Delta\upsilon_{(\upsilon_{asymm.}-\upsilon_{sym})}$ separation of 228 cm^{-1}, which strongly suggest the monodentate linkage of the carboxylate groups. The peak at 3367 cm^{-1} corresponds to the N-H stretching vibration.

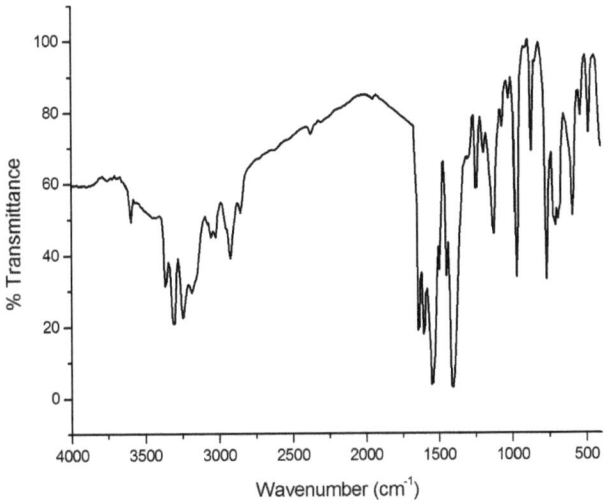

Fig. 5.1 IR spectrum of [Ni$_{0.25}$Co$_{0.75}$Fe$_2$(cin)$_3$(N$_2$H$_4$)$_5$]

5.1.2.3. Thermal analysis

The TG curve in Fig. 5.2 shows that the complex loses weight in three particular steps. The first step which is a minor weight loss step (20%) is related to the removal of the hydrazine molecules between room temperature and 190 °C. The corresponding peak in DSC is observed as an exotherm. The second step is associated with the total decarboxylation of the dehydrazinated complex with a weight loss of 22%, leading to the partial formation of Ni$_{0.25}$Co$_{0.75}$Fe$_2$O$_4$. The last step at 347 °C corresponds to the formation Ni$_{0.25}$Co$_{0.75}$Fe$_2$O$_4$ as the final residue.

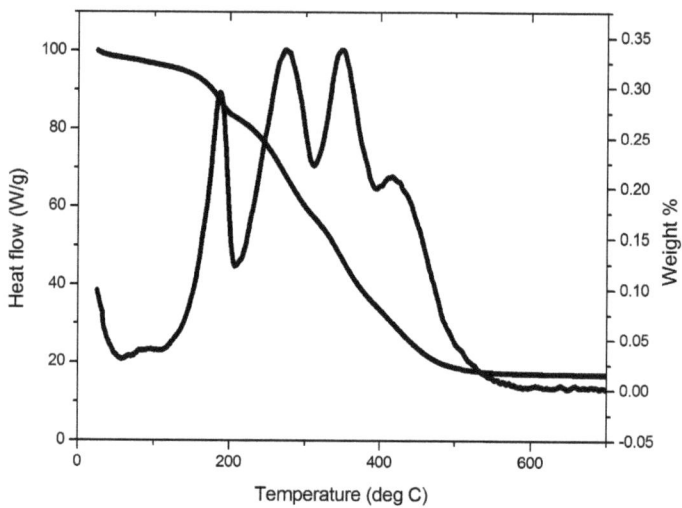

Fig. 5.2 TG-DSC curve of [Ni$_{0.25}$Co$_{0.75}$Fe$_2$(cin)$_3$(N$_2$H$_4$)$_5$]

5.1.2.4. SEM and EDX analysis

From the SEM picture in Fig. 5.3, randomly distributed nano-sized homogenous grains with some agglomerations are noticeable. The EDX spectrum of Ni$_{0.25}$Co$_{0.75}$Fe$_2$O$_4$ nanoparticles is presented in Fig. 5.4, which furnishes the chemical compositional analysis of Ni$_{0.25}$Co$_{0.75}$Fe$_2$O$_4$. The compositional data from the EDX analysis agree well with the theoretically calculated values, indicating the stoichiometry of the sample.

Fig. 5.3 SEM image of $Ni_{0.25}Co_{0.75}Fe_2O_4$

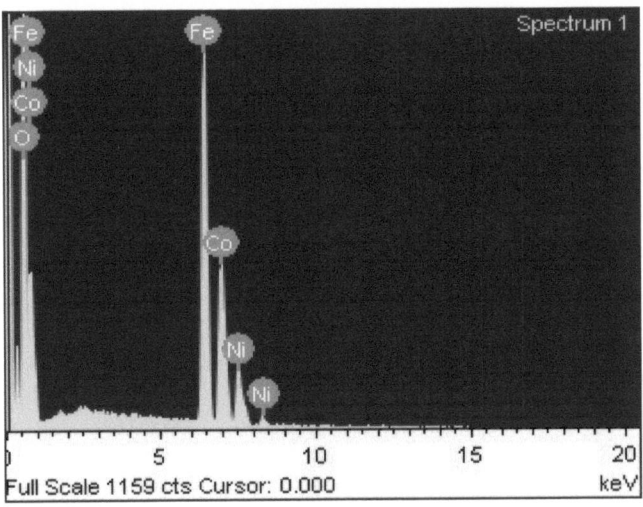

Fig. 5.4 EDX spectrum of $Ni_{0.25}Co_{0.75}Fe_2O_4$

5.2. Synthesis and characterization of [Co$_{0.8}$Zn$_{0.2}$Fe$_2$(cin)$_3$(N$_2$H$_4$)$_4$]

5.2.1. Synthesis of [Co$_{0.8}$Zn$_{0.2}$Fe$_2$(cin)$_3$(N$_2$H$_4$)$_4$]

This complex was prepared by the addition of an aqueous solution (50 mL) of hydrazine hydrate (1.6 mL, 0.02 mol) and cinnamic acid (1.18 g, 0.0079 mol) to the corresponding aqueous solution (50 mL) of cobalt nitrate hexahydrate (0.4656 g, 0.0015 mol), zinc nitrate hexahydrate (0.116 g, 0.0003 mol) and ferrous sulphate heptahydrate (1.11 g, 0.0039 mol). The obtained brown orange precipitate was kept aside for an hour to settle, then filtered and washed several times with distilled water and alcohol followed by diethylether and dried at room temperature.

5.2.2. Characterization of [Co$_{0.8}$Zn$_{0.2}$Fe$_2$(cin)$_3$(N$_2$H$_4$)$_4$]

5.2.2.1. Analytical data

The observed percentage of hydrazine (16.74), zinc (6.01), cobalt (18.92) and iron (47.56) are found to match closely with the calculated values (17.37), (5.54), (19.98) and (47.34) for hydrazine, zinc, cobalt and iron respectively. Based on these values, the chemical formula [Co$_{0.8}$Zn$_{0.2}$Fe$_2$(cin)$_3$(N$_2$H$_4$)$_4$] has been given to the complex.

5.2.2.2. IR spectral analysis

Fig. 5.5 displays the IR spectrum of [Co$_{0.8}$Zn$_{0.2}$Fe$_2$(cin)$_3$(N$_2$H$_4$)$_4$]. A characteristic band at 972 cm^{-1}, assignable to the N-N stretching frequency, explicitly proves the bidentate bridging character of the hydrazine moieties. The asymmetric and symmetric stretching frequencies of the carboxylate ions at 1639 and 1411 cm^{-1}, respectively with the $\Delta \upsilon$ (υ_{asymm}- υ_{sym}) separation of 228 cm^{-1} signify the monodentate interaction of the carboxylate groups. The N-H stretching is observed at 3367 cm^{-1}.

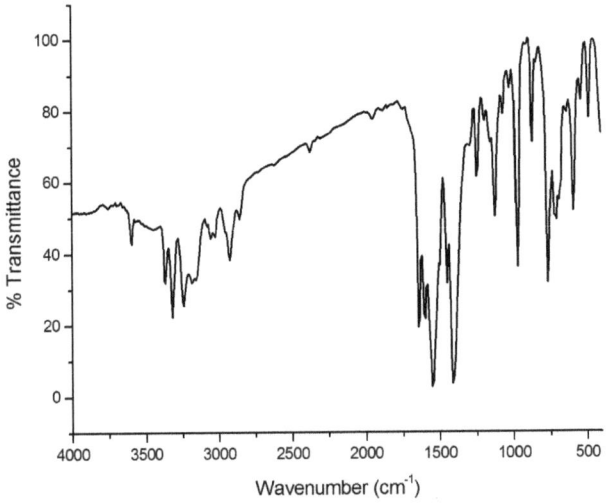

Fig. 5.5 IR spectrum of [Co$_{0.8}$Zn$_{0.2}$Fe$_2$(cin)$_3$(N$_2$H$_4$)$_4$]

5.2.2.3. Thermal analysis

The TG-DSC curve shown in Fig. 5.6 shows four distinct regions of weight loss of the complex. The first step of weight loss at 170 °C corresponds to the removal of the hydrazine molecules with a weight loss of 17%. Accordingly, an exothermic slope exists in the DSC curve. The second step begins at 170 °C and ends at 270 °C involving an exothermic process with a strong DSC peak at 273 °C, with a weight loss of 23.29%, which is attributed to the decarboxylation of the dehydrazinated complex. The third step may be due to the early partial formation of Co$_{0.8}$Zn$_{0.2}$Fe$_2$O$_4$ which is expected to be formed due to the exothermic energy of burning in the previous step. The last step at 517 °C is presumably related to the formation Co$_{0.8}$Zn$_{0.2}$Fe$_2$O$_4$.

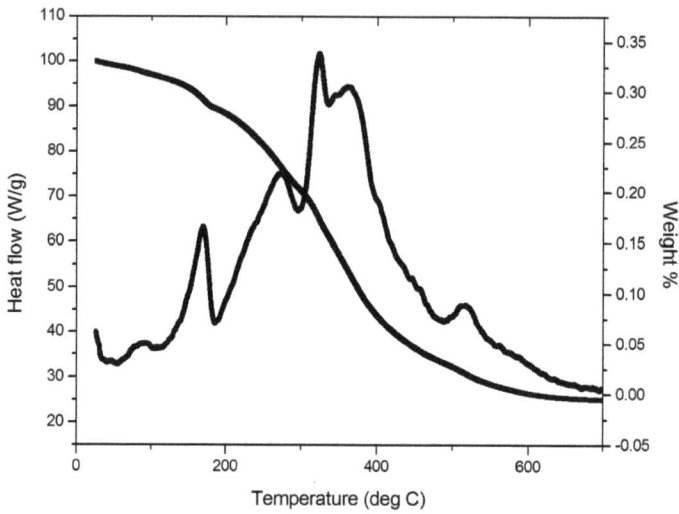

Fig. 5.6 TG-DSC curve of [Co$_{0.8}$Zn$_{0.2}$Fe$_2$(cin)$_3$(N$_2$H$_4$)$_4$]

5.2.2.4. SEM and EDX analysis

The SEM pictures in Fig. 5.7 clearly show that the sample consists of homogenous grains with the presence of a sizable number of agglomerated particles. The grains appear to stick each other and agglomerate in different masses throughout the micrograph.

EDX spectrum of Co$_{0.8}$Zn$_{0.2}$Fe$_2$O$_4$ is presented in Fig. 5.8, which furnishes the chemical compositional analysis of Co$_{0.8}$Zn$_{0.2}$Fe$_2$O$_4$. From the EDX spectrum, it is concluded that the obtained powder consists of the proposed chemical formula, Co$_{0.8}$Zn$_{0.2}$Fe$_2$O$_4$. It is quite clear that the experimentally observed percentages of elements are in a good agreement with those theoretically calculated.

Fig. 5.7 SEM image of $Co_{0.8}Zn_{0.2}Fe_2O_4$

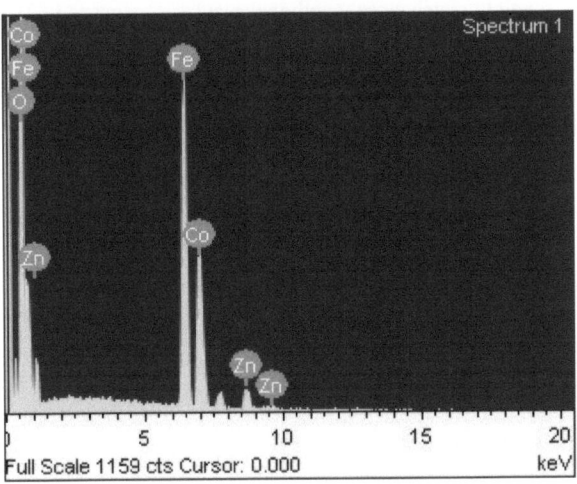

Fig. 5.8 EDX spectrum of $Co_{0.8}Zn_{0.2}Fe_2O_4$

5.3. Synthesis and characterization of [Ni$_{0.8}$Zn$_{0.2}$Fe$_2$(cin)$_3$(N$_2$H$_4$)$_3$]

5.3.1. Synthesis of [Ni$_{0.8}$Zn$_{0.2}$Fe$_2$(cin)$_3$(N$_2$H$_4$)$_3$]

This was prepared by adding an aqueous solution (50 mL) of hydrazine hydrate (1.6 mL, 0.02 mol) and cinnamic acid (1.18 g, 0.0079 mol) to the aqueous solution (50 mL) of nickel nitrate hexahydrate (0.4656 g, 0.0016 mol), zinc nitrate hexahydrate (0.116 g, 0.0003 mol) and ferrous sulphate heptahydrate (1.11 g, 0.0039 mol). The brown orange product formed was kept for an hour, then filtered and washed with water and alcohol repeatedly, followed by diethylether to remove adsorbed impurities. The sample was then air dried and stored in desiccator.

5.3.2. Characterization of [Ni$_{0.8}$Zn$_{0.2}$Fe$_2$(cin)$_3$(N$_2$H$_4$)$_3$]

5.3.2.1. Analytical data

The observed percentage of hydrazine (13.09), nickel (19.43), zinc (5.48) and iron (47.17) are found to match closely with the calculated values (13.62), (19.91), (5.54) and (47.38) for hydrazine, nickel, zinc and iron respectively. Thus the chemical formula [Ni$_{0.8}$Zn$_{0.2}$Fe$_2$(cin)$_3$(N$_2$H$_4$)$_3$] has been assigned to the complex.

5.3.2.2. IR spectral analysis

Fig. 5.9 shows the IR spectrum of [Ni$_{0.8}$Zn$_{0.2}$Fe$_2$(cin)$_3$(N$_2$H$_4$)$_3$], from which one can witness a characteristic band at 972 cm^{-1} due to the N-N stretching frequency, providing a dramatic proof for the bidentate bridging nature of the hydrazine moieties. The asymmetric and symmetric stretching frequencies of the carboxylate ions at 1639 and 1411cm^{-1}, respectively with the Δυ $_{(υ_{asymm}- υ_{sym})}$ separation of 228 cm^{-1}, designate the monodentate linkage of the carboxylate groups. The N-H stretching is observed at 3367 cm^{-1}.

5.3.2.3. Thermal analysis

As can be observed from Fig. 5.10, the weight loss of [Ni$_{0.8}$Zn$_{0.2}$Fe$_2$(cin)$_3$(N$_2$H$_4$)$_3$] can be divided into three regions. In the first one, the sample weight decreases from room temperature to 200 °C and finds a percentage weight loss of 13%, which is expected to be due to the removal of the hydrazine molecules. An obvious exothermic peak is observed on the DSC curve. The second and third steps are ascribed to be the decarboxylation of the dehydrazinated compound, which gives Ni$_{0.8}$Zn$_{0.2}$Fe$_2$O$_4$ as the final residue.

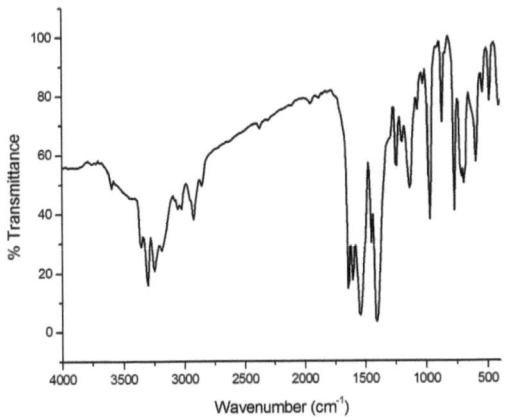

Fig. 5.9 IR spectrum of [Ni$_{0.8}$Zn$_{0.2}$Fe$_2$(cin)$_3$(N$_2$H$_4$)$_3$]

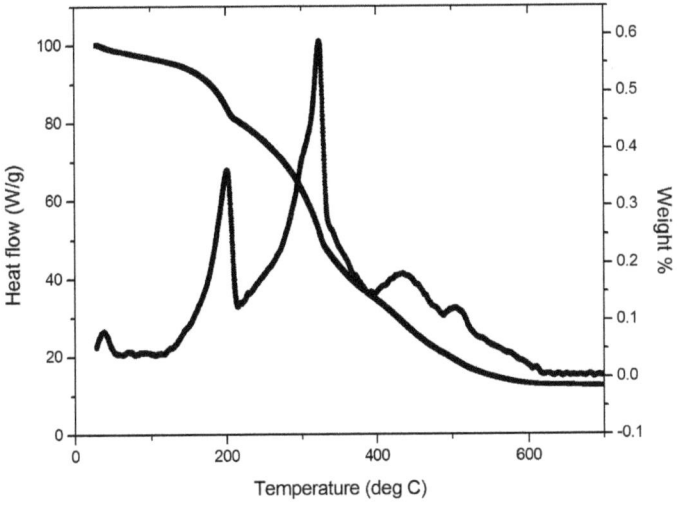

Fig. 5.10 TG-DSC curve of [Ni$_{0.8}$Zn$_{0.2}$Fe$_2$(cin)$_3$(N$_2$H$_4$)$_3$]

5.3.2.4. SEM and EDX analysis

The SEM pictures in Fig. 5.11 clearly show that the powder is mostly formed of spherical grains with the presence of some coalesced particles. EDX spectrum of $Ni_{0.8}Zn_{0.2}Fe_2O_4$ is presented in Fig. 5.12, which furnishes its chemical composition. As expected, the spectrum does not possess any unwanted peaks other than that of Ni, Zn, Fe and O.

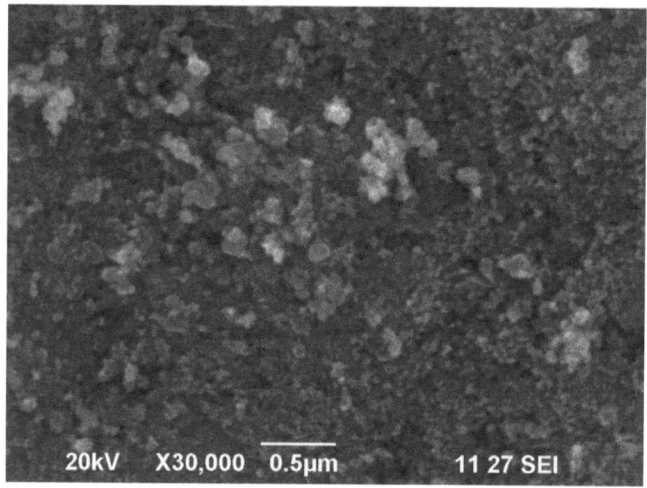

Fig. 5.11 SEM image of $Ni_{0.8}Zn_{0.2}Fe_2O_4$

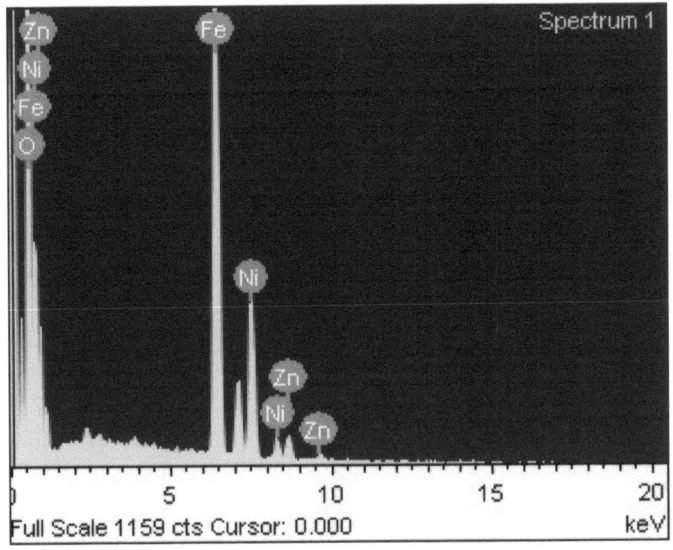

Fig. 5.12 EDX spectrum of $Ni_{0.8}Zn_{0.2}Fe_2O_4$

5.4. Synthesis and characterization of $[Cd_{0.3}Zn_{0.7}Fe_2(cin)_3(N_2H_4)_2]$

5.4.1. Synthesis of $[Cd_{0.3}Zn_{0.7}Fe_2(cin)_3(N_2H_4)_2]$

This was prepared by adding up an aqueous solution (50 mL) of hydrazine hydrate (1.6 mL, 0.02 mol) and cinnamic acid (1.18 g, 0.0079 mol) to the aqueous solution (50 mL) of cadmium nitrate hexahydrate (0.3 g, 0.0009 mol), zinc nitrate hexahydrate (0.7 g, 0.0023 mol) and ferrous sulphate heptahydrate (1.11 g, 0.0039 mol). The brown orange product formed was filtered after an hour, and washed with water, alcohol and diethylether. Then the product was dried in air.

5.4.2. Characterization of $[Cd_{0.3}Zn_{0.7}Fe_2(cin)_3(N_2H_4)_2]$

5.4.2.1. Analytical data

The complex has been assigned the chemical formula $[Cd_{0.3}Zn_{0.7}Fe_2(cin)_3(N_2H_4)_2]$, on the basis of the observed percentage of hydrazine (9.35), cadmium (12.90), zinc (17.31) and iron (42.98) which are found to match closely with the calculated values (9.60), (13.21), (17.92) and (43.76) for hydrazine, cadmium, zinc and iron respectively.

5.4.2.2. IR spectral analysis

The IR spectrum of $[Cd_{0.3}Zn_{0.7}Fe_2(cin)_3(N_2H_4)_2]$ in Fig. 5.13 exhibits a strong band at 972 cm^{-1} due to the N-N stretching frequency, which explicitly proves the bridging nature of the hydrazine moieties. The asymmetric and symmetric stretching frequencies of the carboxylate ions are seen at 1639 and 1411 cm^{-1}, respectively with the $\Delta\upsilon$ (υ_{asymm}- υ_{sym}) separation of 228 cm^{-1}, which indicates the monodentate linkage of the carboxylate groups. The N-H stretching is observed at 3363 cm^{-1}.

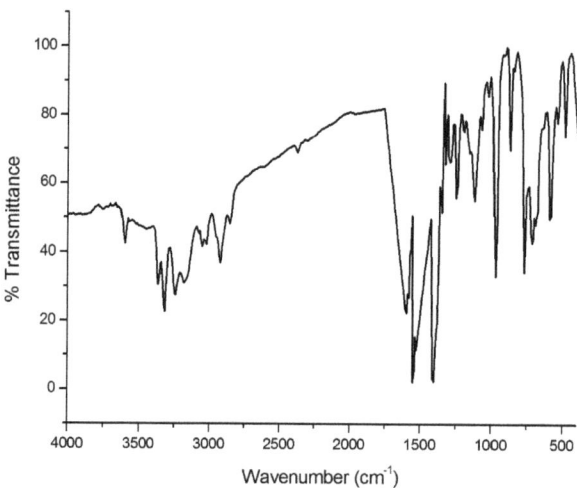

Fig. 5.13 IR spectrum of $[Cd_{0.3}Zn_{0.7}Fe_2(cin)_3(N_2H_4)_2]$

5.4.2.3. Thermal analysis

The TG-DSC curve in Fig. 5.14 shows that the compound loses weight in three particular steps. The first step is the dehydrazination, which is the removal of hydrazine molecules from the complex, taking place between room temperature and 190 °C with a weight loss of 5%. The corresponding peak in DSC is observed as an endotherm. The second and third steps are attributed to the decarboxylation of the dehydrazinated complex, which gives $Cd_{0.3}Zn_{0.7}Fe_2O_4$ as the final residue.

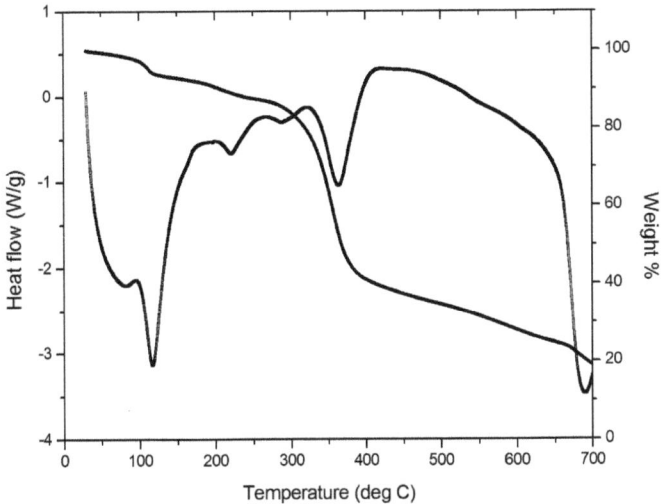

Fig.5.14 TG-DSC curve of [Cd$_{0.3}$Zn$_{0.7}$Fe$_2$(cin)$_3$(N$_2$H$_4$)$_2$]

5.4.2.4. SEM and EDX analysis

The SEM picture in Fig. 5.15 clearly shows randomly distributed grains with fairly uniform size. A large agglomeration of particles is noticeable which may be due to their magnetic interaction.

EDX spectrum of Cd$_{0.3}$Zn$_{0.7}$Fe$_2$O$_4$ is presented in Fig. 5.16. The actual chemical composition of Cd$_{0.3}$Zn$_{0.7}$Fe$_2$O$_4$ is known from the spectrum, which consists of the elements Cd, Zn, Fe and O. No other impurities are noticeable.

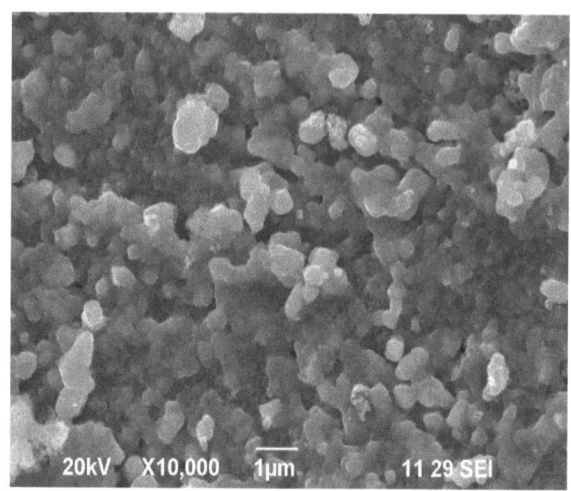

Fig.5.15 SEM image of $Cd_{0.3}Zn_{0.7}Fe_2O_4$

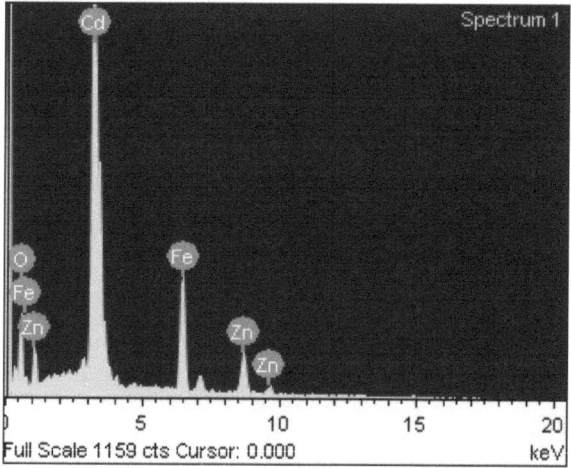

Fig. 5.16 EDX spectrum of $Cd_{0.3}Zn_{0.7}Fe_2O_4$

Conclusions

Hetero trimetallic hydrazine cinnamates were successfully synthesised from the respective metal salts, hydrazine and cinnamic acid, and were characterized by analytical data, IR spectroscopy and TG-DSC. The complexes yielded their corresponding mixed metal oxides (ferrites) as the decomposition product. These exhibited various morphologies which were evaluated by SEM. The EDX analysis was also carried out for all the samples in order to find out their actual chemical composition.

CHAPTER VI
SUMMARY, CONCLUSIONS AND FURTHER SCOPE

This chapter describes the salient features of the work undertaken in this thesis and discusses the possible avenues for future work.

6.1 Summary of the work

The work in this thesis has been devoted towards exploitation of the synthesis of transition metal hydrazine cinnamates using transition metals salts, hydrazine hydrate and cinnamic acid. An introduction about coordination complexes, chemistry of hydrazine, cinnamic acid and transition metals, their applications, various metal hydrazine carboxylates are presented in detail in Chapter I. The scope and objectives of the study are also discussed.

The specifications of all the materials used in the study and the details of the different experimental techniques employed in this study are elaborated in Chapter II.

Chapter III illustrates the synthesis of the complex $[M(cin)_2(N_2H_4)_2]$, (M= Ni, Co, Zn or Cd) and its characterisation by analytical data, IR spectroscopy and Thermogravimetry - Differential Thermal Analysis (TG-DTA). The bidentate bridging nature of the hydrazine ligand and the monodentate coordination of the carboxylate groups to the metal ions in the complexes are known form the IR spectral data. The thermal decomposition of the prepared complexes has led to the formation of the corresponding metal oxides as confirmed by IR spectra, TG-DTA and EDX. The morphology of the obtained oxide materials has been studied by SEM.

Chapter IV details the synthesis of hetero bimetallic hydrazine cinnamates. These complexes have been prepared form their corresponding metal salts, cinnamic acid and hydrazine hydrate. Their characterization has been done by analytical analysis, IR spectral study and TG-DTA. Thermal decomposition of the complexes yielded their respective mixed metal oxides, as indicated by their EDX spectra.

Chapter V emphasises the synthesis and characterisation of hetero trimetallic hydrazine cinnamates, which have been obtained from their respective metal salts, cinnamic acid and hydrazine hydrate. The complexes have been characterised by analytical analysis, IR spectroscopy and thermogravimetry. The thermal decomposition of the prepared complexes has led to the formation of their corresponding mixed metal oxides, which are characterised by SEM-EDX.

6.2 Conclusions

The investigations and observations made on the syntheses and characterisation of of metal and mixed metal hydrazine cinnamates lead to the following conclusions.

In all the complexes prepared from their respective metals, hydrazine hydrate and cinnamic acid by a simple precipitation method, the hydrazine moieties and the carboxylate groups were found to coordinate to the metal in a bridging bidentate and monodentate fashions respectively. All the complexes thermally decomposed through two or three steps, to form their respective oxides, as confirmed by TG-DTA/TG-DSC. The calculated analytical data and the theoretical one were found to agree well with each other. A possible structure was also proposed for the complexes.

The complex $[M(cin)_2(N_2H_4)_2]$ (M=Co, Ni, Zn or Cd), on thermal decomposition produced their respective oxides (Co_3O_4, ZnO and CdO) except $[Ni(cin)_2(N_2H_4)_2]$, which yielded metallic Ni as the final residue. This was confirmed by TG-DTA and EDX. SEM images reported the morphology of the samples. CdO nanoparticles showed a rock-candy like morphology, unlike the other three, which showed sphere-shaped morphologies.

The synthesised complex $[MFe_2(cin)_3(N_2H_4)_3]$ (M=Co, Ni, Zn or Cd) on thermal decomposition, yielded their corresponding mixed metal oxide (ferrite) MFe_2O_4. SEM results indicated more or less spherical morphology of particles.

Thermal decomposition of hetero trimetallic hydrazine cinnamates also resulted in the formation of mixed metal oxides (ferrites). This was confirmed by EDX. Sphere like particles were seen from SEM images of all samples. A considerable extent of agglomeration was also noticed which may be attributed to their magnetic interaction.

6.3 Further scope

A tremendous surge in research on transition metal complexes has been observed in the last few years since these materials are to shape the future technology. As all the complexes produced metal oxides and mixed metal oxides (ferrites) as their thermal decomposition product, this could be made use of in synthesising them in their nanoscale. Thus, they can serve as precursors for nanosized metal oxides and ferrites.

Spinel ferrites are widely used as magnetic materials, gas sensor, catalysts, photocatalysts and absorbent materials. Synthesis, development and study of different properties of the ferrite nanomaterials using various techniques has become an interesting area of research. The research on spinel ferrites received a major boost in recent years due to the availability of new and sophisticated techniques for the synthesis and characterization of nanoparticles. Since there is always a scope for improvement and widening the area of application, good amount of work can be taken up on these materials. Similar study of these materials, synthesized using microwave decomposition could be interesting. This method could be compared with the materials prepared using the present thermal decomposition method. They can also be investigated for their dielectric and catalytic properties. The materials can also be studied more for their applications as sensors for toxic gases in the environment. Also these materials can be reviewed for their potential use in medical applications, especially with regard to drug delivery and hyperthermia. Thus there exists ample scope to carry out further studies on the applications of these ferrite materials.

REFERENCES

[1] G.L. Miessler and D.A. Tarr, Inorganic Chemistry, 3rd Edition Pearson Prentice Hall, 2004.
[2] T. Curtius, T. Prakt. Chem. 39 (1889)107.
[3] F. Raschig, Ber.D.Chem.Ges. 40 (1907) 4587.
[4] Riegel, Emil Raymond. "Hydrazine". Riegel's Handbook of Industrial Chemistry. (1992) 192.
[5] Chemistry of Petrochemical Processes, 2nd edition, Gulf Publishing Company, 1994 (2000) 148.
[6] "Hydrazine: Chemical product info". chemindustry.ru, Retrieved 2007-01-08.
[7] L.F. Audrieth and B.A. Ogg, "The chemistry of Hydrazine", John Wiley, New York, 1951.
[8] C.C. Clark, "Hydrazine", Mathieson Chem., Corp. Baltimore Mol. 1953.
[9] F. Bottomely, Quart. Rev., Chem. Soc. 24 (1970) 617.
[10] E.W. Schmidt, "Hydrazine and its derivatives – preparation, properties and applications", Wiley Interscience, New York, 1984.
[11] J.C. Decius and D.P. Pearson, *J. Am. Chem. Soc.*, 75 (1953) 2436.
[12] G. Pannetier, F. Margineanu, A. Dereigne and R. Bonnaire, *Bull. Soc. Chim. Fr.*, 7 (1972) 2623.
[13] R.G. Snyder and J.C. Decius, *Spectrochim. Acta.*,13 (1959) 280.
[14] P.W.M. Jacobs and A. Russel-Jones, *Can. J. Chem.*,44 (1966) 2435.
[15] R. Savoie and M. Guay, *Can. J. Chem.*, 53 (1975) 1387.
[16] A.V.R. Warrier and P.S. Narayanan, *Int. J. Pure Appl. Phys.*, 5 (1967) 216.
[17] P. Glavic and J. Slivnik, *Monatsh. Chem.*, 98 (1967) 1878.
[18] J. Lundgren, J. De Villepin and A. Novak, *Chem. Phys. Lett.*, 3 (1969) 84.
[19] P. Glavic and D. Hadzi, *Spectrochim. Acta.*, 28A (1972) 1963.
[20] F.G. Baglin, S.F. Bush and J.R. Durig, *J. Chem. Phys.*, 47 (1967) 2104.
[21] P.A. Gigurere and I.D. Liu, *J. Chem. Phys.*, 20 (1952) 136.
[22] A. Braibanti, F. Dallavalle, M.A. Pellinghelli and E. Laporati, *Inorg. Chem.*, 7 (1968) 1430.

[23] L. Sacconi and A. Sabatini, *J. Inorg. Nucl. Chem.*, 25 (1963) 1389.

[24] Z. Mielke and H. Ratajczak, *J. Mol. Struct.*, 19 (1973) 751.

[25] J.R. Durig, S.F. Bush and E.E. Mercer, *J. Chem. Phys.*, 44 (1966) 4238.

[26] D.N. Satyanarayana and D. Nicholls, *Spectrochim. Act.*, 34A (1978) 268.

[27] Vieira, R.; C. Pham-Huu, N. Keller, M. J. Ledoux, Chemical Communications. 9 (2002) 954.

[28] Chen, Xiaowei; et al. Catalysis Letters. 79 (2002) 21.

[29] V.M.S. Verenkar, K.S. Rane, P.Y. Sawant, J Mater Sci Mater Electr.10 (1999) 133.

[30] K.S. Rane, V.M.S. Verenkar, R.M. Pednekar, P.Y. Sawant, J Mater SciMater Electr. 10 (1999) 121.

[31] K.S. Rane, V.M.S. Verenkar, Bull. Mater. Sci. 24 (2001) 39.

[32] A. More, V.M.S. Verenkar, Inorganic materials: recent advances. New Delhi: Narosa Publishing House, (2006) 230.

[33] U.B. Gawas, V.M.S. Verenkar, S. C. Mojumdar, J. Therm. Anal. Calorim. 104 (2011) 879.

[34] Reema A. Porob, Sitara Z. Khan , S. C. Mojumdar, V. M. S. Verenkar,J. Therm. Anal. Calorim. 86 (2006) 605.

[35] R.A. Porob, S.Y. Sawant, K.R. Kannan, V.M.S. Verenkar, Proceedings of the 14th national symposium on thermal analysis, Baroda. Mumbai: Indian Thermal Analysis Society. (2004) 335.

[36] R.A. Porob, S.Z. Khan, S.C. Mojumdar, V.M.S Verenkar, 15th Canadian thermal analysis society's annual workshop and exhibition. Boucherville, Quebec, Canada: National Research Council Canada. 2005.

[37] S.Y. Sawant, S.C. Mojumdar, V.M.S. Verenkar, 16th Canadian thermal analysis society's annual workshop and exhibition, Canada. 2006.

[38] G.M. Sail, V.M.S. Verenkar, 3rd national symposium in chemistry. Chandigarh, India: Punjab University, (2001) 133.

[39] G.V. Karamanov, Hydrazine derivatives, CA82, 26145.

[40] S. Kondo, J. Am. Chem. Soc. 93 (1971) 6305.

[41] F.R. Quinn, J.S. Driscoll and C. Hansch, J. Med. Chem. 18 (1975) 332.

[42] J. Druey and B.H. Ringier, Helv. Chim. Acta. 34 (1951) 195.

[43] I. Abdelmoty, V. Buchholz Dil, C. Guo, K. Kowitz, V. Enkelmann, G. Weger, and G.M. Foxman, Cryst. Growth. Des., 5 (2005) 2210.

[44] Budavari, Susan, [Ed.], The Merck Index: An Encyclopedia of Chemicals, Drugs, and Biologicals, 12th Edition, Merck, 1996.

[45] D. Garbe "Cinnamic Acid" in Ullmann's Encyclopedia of Industrial Chemistry, Wiley-VCH, Weinheim, 2000.

[46] J.A. Hoskins, J. Appl. Toxicol., 4 (1984) 283.

[47] K.V. Thimann, M.B. Wilkinson (Ed.), McGraw-Hill, London, 2, 1969.

[48] H.S. Chung and J.C. Shin, Food Chem.,104 (2007) 1670.

[49] P. De, M. Baltas and F. Bedos-Belval, Curr. Med. Chem., 18 (2011) 1672.

[50] T. Ernawati, Y. Anita, P.D. Lotulung and M. Hanafi, J. Appl. Pharm. Sci., 4 (2014) 092.

[51] S. Naz, S. Ahmad, S.A. Rasool, S.A. Sayeed, and R. Siddiqi, Microbiol. Res., 161 (2006) 43.

[52] S.A. Carvalho, E.F. da Silva, M.V.N. de Souza, M.C.S. Lourenço and F.R. Vicente, Bioorg. Med. Chem. Lett., 18 (2008) 538.

[53] H.B. Zhou, S.Y. Dong, C.X. Zhou, L.H. Hu, Y.H. Wu, H.B. Li, J.X. Gong, L.L. Sun, X.M. Wu, H. Bai, B.T. Fan, X.J. Hao, J. Stöckigt and Y. Zhao, Bioorg. Med. Chem., 14 (2006) 2060.

[54] Jensen, B. William, Journal of Chemical Education. 80 (2003) 952.

[55] C. R. Bury, J. Amer. Chem. Soc. 43 (1921) 1602.

[56] Enghag, "Cobalt". Encyclopedia of the elements: technical data, history, processing, applications. (2004) 667.

[57] V.S.R. Murthy, Structure And Properties Of Engineering Materials. (2003) 381.

[58] Celozzi, Salvatore, Araneo, Rodolfo, Lovat, Giampiero, Electromagnetic Shielding. (2008) 27.

[59] B. Lee, R. Alsenz, A. Ignatiev, M. Van Hove, Physical Review B. 17 (1978) 1510.

[60] "Properties and Facts for Cobalt". American Elements. Retrieved 2008-09-19.

[61] Cobalt, Centre d'Information du Cobalt, Brussels. (1966) 45.

[62] A.F. Holleman, E. Wiberg, N. Wiberg, "Cobalt" (in German). Lehrbuch der Anorganischen Chemie, (2007) 1146.

[63] C.E. Housecroft, A.G. Sharpe, Inorganic Chemistry. 3 (2008) 722.

[64] Shedd, B. Kim. "Mineral Yearbook 2006: Cobalt". United States Geological Survey.. Retrieved 2008-10-26.

[65] Shedd, B. Kim. "Commodity Report 2008: Cobalt". United States Geological Survey. Retrieved 2008-10-26.

[66] R. Michel, M. Nolte, M. Reich, F. Löer, "Systemic effects of implanted prostheses made of cobalt-chromium alloys". Archives of Orthopaedic and Trauma Surgery 110 (1991) 61.

[67] F.E. Luborsky, L.I. Mendelsohn, T.O. Paine, "Reproducing the Properties of Alnico Permanent Magnet Alloys with Elongated Single-Domain Cobalt-Iron Particles". Journal Applied Physics. 28 (1957) 344.

[68] M. Hawkins, "Why we need cobalt". Applied Earth Science: Transactions of the Institution of Mining & Metallurgy, Section B. 110 (2001) 66.

[69] R.D. Armstrong, G.W.D. Briggs, E.A. Charles, Journal of Applied Electrochemistry. 18 (1988) 215.

[70] P. Zhang, Journal of Power Sources. 77 (1999) 116.

[71] Khodakov, Y. Andrei, Chu, Wei, Fongarland, Pascal, Chemical Review. 107 (2007) 1692.

[72] Muhlethaler, Bruno, Thissen, Jean, Muhlethaler, Bruno, "Smalt". Studies in Conservation. 14 (1969) 47.

[73] A.F. Gehlen, "Ueber die Bereitung einer blauen Farbe aus Kobalt, die eben so schön ist wie Ultramarin. Vom Bürger Thenard". Neues allgemeines Journal der Chemie, Band 2 (H. Frölich.). (German translation from L. J. Thénard; Journal des Mines; Brumaire 12 (1803) 128.

[74] H.J. Witteveen, E.F. Farnau, Industrial & Engineering Chemistry. 13 (1921) 1061.

[75] S. Venetskii, "The charge of the guns of peace". Metallurgist. 14 (1970) 334.

[76] Kittel, Charles, Introduction to Solid State Physics. Wiley. (1996) 449.

[77] Greenwood, N. Norman. Earnshaw, Alan, Chemistry of the Elements (2nd ed.). Butterworth–Heinemann. (1997).

[78] Kuck, H. Peter H, "Mineral Commodity Summaries 2012: Nickel". United States Geological Survey. Retrieved 2008-11-19

[79] Kuck, H. Peter, "Mineral Yearbook 2006: Nickel". United States Geological Survey.

[80] Davis, R. Joseph, "Uses of Nickel". ASM Specialty Handbook: Nickel, Cobalt, and Their Alloys. ASM International. (2000) 7.

[81] Kharton, V. Vladislav, Solid State Electrochemistry II: Electrodes, Interfaces and Ceramic Membranes. Wiley-VCH. (2012) 166.

[82] F. Bidault, D.J.L. Brett, P.H. Middleton, N.P. Brandon, "A New Cathode Design for Alkaline Fuel Cells(AFCs)". Imperial College London.

[83] UCLA – Magnetostrictive Materials Overview. Aml.seas.ucla.edu. Retrieved on 2012-01-09.

[84] R.F. Cheburaeva, I.N. Chaporova, T.I. Krasina, Soviet Powder Metallurgy and Metal Ceramics. 31 (1992) 423.

[85] Astrid Sigel, Helmut Sigel, Roland K. O. Sigel, Astrid Sigel, Helmut Sigel, Roland K. O. Sigel. Nickel and Its Surprising Impact in Nature. Metal Ions in Life Sciences. Wiley 2 (2008).

[86] A.F. Holleman, E. Wiberg, Wiberg, Nils "Cadmium". Lehrbuch der Anorganischen Chemie, Walter de Gruyter. (1985) 1056.

[87] "Case Studies in Environmental Medicine (CSEM) Cadmium". Agency for Toxic Substances and Disease Registry. Archived from the original on 2011-06-06.

[88] Scoullos, J. Michael, Vonkeman, H. Gerrit, Thornton, Iain; Makuch, Zen. Mercury, Cadmium, Lead: Handbook for Sustainable Heavy Metals Policy and Regulation. Springer (2001).

[89] Spelter. Encyclo. Retrieved 2009-08-01.

[90] Scoffern, John (1861). The Useful Metals and Their Alloys. Houlston and Wright. (2009) 591.

[91] R. S. Lehto,"Zinc, The Encyclopedia of the Chemical Elements , Clifford A. Hampel. New York: Reinhold Book Corporation, (1968) 822.

[92] "Zinc Metal Properties". Americal galvanizers association. Retrieved 2009-02-15.

[93] Ritchie, Rob. Chemistry Letts and Lonsdale. (2004) 71.

[94] Burgess, John. Metal ions in solution. New York: Ellis Horwood. (1978) 147.

[95] Holleman, F. Arnold, Wiberg, Egon, Wiberg, Nils, "Zink" (in German). Lehrbuch der Anorganischen Chemie. Walter de Gruyter. (1985) 1034.

[96] Brady, E. James, Humiston, E. Gerard, Heikkinen, Henry. General Chemistry: Principles and Structure. John Wiley & Sons. (1983) 671.

[97] "Zinc: World Mine Production (zinc content of concentrate) by Country". Minerals Yearbook: Zinc. Washington, D.C.: United States Geological Survey. February 2010.

[98] M. Bounoughaz, E. Salhi, K. Benzine, E. Ghali, F. Dalard, Journal of Materials Science. 38 (2003) 1139.

[99] Besenhard, Jürgen O. Handbook of Battery Materials. Wiley-VCH. Retrieved 2008-10-08.

[100] J.P. Wiaux, J.P. Waefler, Journal of Power Sources. 57 (1995) 61.

[101] T. Culter, "A design guide for rechargeable zinc-air battery technology". Southcon/96. Conference Record. (1996) 616.

[102] Whartman, Jonathan; Brown, Ian. "Zinc Air Battery-Battery Hybrid for Powering Electric Scooters and Electric Buses". The 15th International Electric Vehicle Symposium. Retrieved 2008-10-08.

[103] J.F. Cooper, D. Fleming, D. Hargrove, R. Koopman, K. Peterman, K. "A refuelable zinc/air battery for fleet electric vehicle propulsion". Society of Automotive Engineers future transportation technology conference and exposition. Retrieved 2008-10-08.

[104] Eastern Alloys contributors. "Diecasting Alloys". Maybrook, NY: Eastern Alloys. Retrieved 2009-01-19.

[105] Davies, Geoff. Materials for automobile bodies. Butterworth-Heinemann. (2003) 157.

[106] Blew, Joseph Oscar. "Wood preservatives". Department of Agriculture, Forest Service, Forest Products Laboratory. (1953).

[107] Paschotta, Rüdiger. Encyclopedia of Laser Physics and Technology. Wiley-VCH. (2008) 798.

[108] I.K. Konstantinou, T.A. Albanis, Environment International. 30 (2004) 235.

[109] Boudreaux, A. Kevin. "Zinc + Sulfur". Angelo State University. (2008).

[110] Milbury, E. Paul, Richer, C. Alice. Understanding the Antioxidant Controversy: Scrutinizing the "fountain of Youth". Greenwood Publishing Group (2008) 99.

[111] S. Roldán, E.G. Winkel, D. Herrera, M. Sanz, A.J. Van Winkelhoff, Journal of Clinical Periodontology. 30 (2003) 427.

[112] R. Marks, A.D. Pearse, A.P. Walker, British Journal of Dermatology 112 (1985) 415.

[113] T.J. McCarthy, J.J. Zeelie, D.J. Krause, Clinical Pharmacology & Therapeutics (American Society for Clinical Pharmacology and Therapeutics) 17 (1992) 5.

[114] Nam, Wonwoo, Accounts of Chemical Research 40 (2007) 522.

[115] Holleman, F. Arnold, Wiberg, Egon, Wiberg, Nils, "Iron" (in German). Lehrbuch der Anorganischen Chemie (91–100 ed.). Walter de Gruyter. (1985) 1125.

[116] Reiff, William Michael, Long, J. Gary, Mössbauer spectroscopy applied to inorganic chemistry. Springer (1984) 245.

[117] Camp, James McIntyre; Francis, Charles Blaine, The Making, Shaping and Treating of Steel. Pittsburgh: Carnegie Steel Company. (1920) 173.

[118] Wildermuth, Egon, Stark, Hans, Friedrich, Gabriele, Ebenhöch, Franz Ludwig, Kühborth, Brigitte, Silver, Jack, Rituper, Rafael, Iron Compounds. (2000).

[119] M.S. Bans, D.C. Bradley Can., J. Chem. 40 (1962) 1351.

[120] D. Sellmann, H. Friedrich, F. Knoch, Z. Naturforch, Inorg. Chem. 49b (1994) 660.

[121] D. Sellmann, W. Soglowek, F. Knoch, G. Ritler, J. Dengler, Inorg. Chem. 31 (1992) 3711.

[122] B.T. Heaton, C. Jacob, P. Page, Coord. Chem. Rev. 154 (1996) 193.

[123] A. Ferrari, A. Braibanti, G. Bigliardi, A.M. Lanfredi, Acta Cryst. 19 (1965) 548.

[124] D.T. Cromer, A.C. Larson, R.B. Roof., Acta Cryst. 20 (1966) 279.

[125] M. Gustafsson, A. Fischer, A. Ilyukhin, M. Maliarik, P. Nordblad, Inorg. Chem. 49 (2010) 5359.

[126] S. Devipriya, N. Arunadevi, S. Vairam, Journal of Chemistry. (2013) 10.

[127] R.Ragul, B.N. Sivashankar, J. Chem Crystallogr. 42 (2012) 533.

[128] R. Ragul, B. N. Sivashankar, Synth reac. metal-organic. nano-metal. (2013) 382.

[129] S. Vairam, T. Premkumar, S. Govindarajan, J. Therm. Anal. 101 (2009) 979.

[130] R.L. Hinman, J. Org. Chem. 23 (1958)1587.

[131] J. Chatt, G.L. Leigh, R.J. Paske, J. Chem. Soc. A. (1969) 854.

[132] J. Chatt, N.P. Johnson, B.L. Shaw, J.Chem. Soc. A (1964) 2508.

[133] D.W. Bisacchi, H. Goldwhite, J. Inorg. Nucl. Chem. 32 (1960) 965.

[134] K.C. Patil, C. Nesamani, V.R. Pai Verneker, Synth. React. Inorg. Met. Org. Chem. 12 (1982) 383.

[135] K.C. Patil, D.Gajapathy, K. Kishore, Thermochim. Acta. 52 (1982) 113.

[136] K.C. Patil, C. Nesamani, V.R. Pai Vernekar, Polyhedron. 1 (1982) 421.

[137] D. Gajapathy, K.C. Patil, V.R. Pai Vernekar, Mat. Res. Bull. 17 (1982) 29.

[138] D. Gajapathy, "Studies on Metal-oxalate-hydrazine system", Ph.D., Thesis, Indian Institute of Science, Bangalore, India (1982).

[139] K.C. Patil, S. Govindarajan, H. Manohar. Synth. React. Inorg. Met. Org. Chem. 11(1981)245.

[140] A. Ferrari, A. Braibanti, G. Bigliardi, A.M. Lanrnedi, Gazz. Chim. Ital. 93 (1963) 937.

[141] G.B. Kauffmann, N. Sugisaka, Z. Anorg. Chem. 344 (1966) 92.

[142] A.K. Srivastava, A.L. Varshney, P.C. Jain, J. Inrog. Nucl. Chem. 42 (1980) 47.

[143] P. Ray, P.V. Sarkar, J. Chem. Soc. 117 (1920) 321.

[144] J.E. House, A.J. Vandenbrook, Thermochim. Acta. 161 (1990) 85.

[145] J. Macek, G. Baric, B. Novosel, A. Rahten, J. Therm. Anal. 36 (1990) 695.

[146] A. Anagnostopouls, D. Nicholls, J. Reed. Inorg. Chim. Acta. 32 (1979) L17.

[147] R.C. Aggarwal, V. Chandrasekar, Indian J. Chem. 17A (1979) 361.

[148] P. Ravindranathan, K.C. Patil, Thermochim. Acta. 71 (1983) 53.

[149] G.V. Mahesh, K.C. Patil. Thermochim. Acta. 99 (1986) 153.

[150] A. Anagnostopouls, D. Nicholls, J. Inorg. Nucl. Chem. 38 (1976) 1615.

[151] V.T. Athavale, C.S. Padmanabha Iyer, J. Inorg, Nucl. Chem. 29 (1967) 1003.

[152] A. Ferrari, A. Braibanti, A.M. Lanfredi, Ann. Chim. (Rome), 48 (1958) 1238.

[153] B.N. Sivasankar, S. Govindarajan, Thermochim. Acta. 244 (1994) 235.

[154] P.V. Gogorishvili, M.V. Karkarasvili, L.G. Tsitsishvili, Zh. Neorg. Khim. 2 (1957) 532.

[155] K.C. Patil, R. Soundararajan, V.R. Pai Verneker, Proc. Indian Acad. Sci. (Chem. Sci.), 88A (1979) 211.

[156] A. Braibanti, G. Bigliardi, R.C. Padovani, Gazzetta, 95 (1965) 877.

[157] A. Ferrari, A. Braibanti, G. Bigliardi, A.M. Lanfredi, A. Tiripicchio, Nature. 211 (1966) 1174.

[158] P.V. Gogorishvili, T.M. Khonelidze, Zh. Neorg. Khim., 5 (1961) 861.

[159] A. Braibanti, G. Bigliardi, R.C. Padovani, Ateneo Parmense, Sez. II 1 (1965) 75: CA. 65, 16164.

[160] M.K. Guseinova, M.A. Porai-Koshits, P.V. Goroishvili, A.S. Antsyshkina, Dokl. Akad. Nauk. SSSR, 169 (1966) 577.

[161] A. Braibanti, F. Dallavalle, A. Tiripicchio, Ric. Sci. 36 (1966) 1156.

[162] D. Hansel, F. Sevsek, J. Solid State Chem. 28 (1979) 385.

[163] A. Braibanti, A.M. Lanfredi, A. Tiripicchio, Z. Crystallogr. 124 (1967) 335.

[164] J. Slivnik, A. Rihar, B. Sedej, Monatsh. Chem. 98 (1967) 200.

[165] H. Funk, A. Eichoff, G. Giesder, Omagiu Raluca Ripan, Acad. Rep. Soc. Romania, (1966) 245.

[166] J. Macek, A. Rahten, Thermochim. Acta, 144 (1989) 257; 224 (1993) 217.

[167] L. Golic, J. Slivnik, M. Levstek, A. Rihar, Monatsh, Chem. 99 (1968) 289.

[168] K.C. Patil, R. Soundararajan, E.P. Goldberg, Synth, React. Inorg. Met. – Org. Chem. 13 (1983) 29.

[169] A. Braibanti, A.M. Lanfredi, A. Tiripicchio, F. Bigoli, Acta Cryst. B25 (1969)100.

[170] A. Braibanti, A. Tiripicchio, A.M. Lanfredi, M. Camellini, Acta Cryst. 23 (1967) 248.

[171] S. Sundar Manoharan, K.C. Patil, Proc. Indian Acad. Sci. (Chem. Sci.), 101 (1989) 377.

[172] D.A. Edwards, D. Thompsett, J.M. Bellerby, J. Chem. Soc. Dalton Trans. (1992) 1761.

[173] B.N.Sivasankar, L.R.Sharmila, Jour. Therm. Anal. Cal. 73(2003)271-283

[174] N.R.S. Kumar, M. Nethaji, K.C. Patil, J. Chem. Soc. Dalton Trans. (1991) 1251.

[175] I.N. Polyakova, G.A. Seisenbaera, Koord. Khim. 3 (1991) 17.

[176] P. Glavic, J. Slivnik, J. Inorg. Nucl. Chem. 32 (1970) 2939.

[177] W. Brzyska, A. Goral, Ann. Univ. Mariae Curie-Sklodowska, Sect. AA, 29-30 (1975) 343.

[178] A. More, V.M.S. Vernekar, S.C. Mojumar, J. Therm. Anal. Cal. 94 (2008) 63.

[179] U. B. Gawas, S. C. Mojumdar, V. M. S. Verenkar, J. Therm. Anal. Cal. 100 (2010) 867.

[180] P. Ravindranathan, G.V. Mahesh and K.C. Patil, J. Solid State Chem. 66 (1987) 20.

[181] L. Vikram, B.N. Sivasankar, Thermochimica Acta. 108 (2007) 865.

[182] K.C. Patil, S. Govindarajan, R. Soundararajan, V.R. Pai Verneker, Proc. Indian Acad. Sci. (Chem. Sci.), 90 (1981) 421.

[183] R.Ya. Aliev, A.D. Kuliev, N.G. Kluchnikov, Zh. Neorg. Khim. 23 (1978) 2239.

[184] K.P. Mamedov, R.A. Aliev, Z.I. Suleimanov, A.D. Kuliev, Zh. Fiz. Khim. 47 (1973) 696.

[185] T.M. Zhdanovskikh, E.I. Krylov, V.A. Sharov, Zh. Neorg. Khim. 24 (1979) 2730.

[186] R. Tsuchiya, M. Yanemura, A. Vehara, E. Kyano, Bull. chem. Soc. Japan. 47 (1974) 660.

[187] P. Glavic, J. Slivnik, A. Bole, J. Inorg. Nucl. Chem. 39 (1977) 259: 41 (1979) 2488; 42 (1980) 617; 42 (1980) 1781.

[188] B. N. Sivasankar, J. R. Sharmila, L. Ragunath Synth. React. Inorg. Met-org. Chem. 34 (2004) 10.

[189] B.N. Sivasankar, S. Govindarajan, Synth. React. Inorg. Met-org. Chem. 25 (1995) 127.

[190] B.N. Sivasankar, S. Govindarajan, Materials Research Bulletin. 31 (1996) 47.

[191] A. More, V.M.S. Verenkar, S.C. Mojumdar, J. Therm. Anal and Calorimetry. 94 (2008) 63.

[192] D. Gajapathy, K.C. Patil, Materials chem & phy. 9 (1983) 423.

[193] K.C. Patil, D. Gajapathy, V.R. Pai Verneker, J. Mat. Sci. Lett. 2 (1983) 272.

[194] J.M. Heintz, J.C. Bernier, J. Mat. Sci. 21 (1986) 1569.

[195] B.N. Sivasankar, "Studies on metal hydrazine carboxylates" Ph.D Thesis, Bharathiar University, Coimbatore, India, 1994.

[196] J. Macek, A. Rahten, J. Slivnik, in D. Dollimore (Ed.) Proc. 1st European Symposium on Thermal Analysis, Salford, Heyden, London (1976) 161.

[197] B. Novosel, J. Macek, V. Ivancevic, J. Therm. Anal. 40 (1993) 427.

[198] J. Macek, D. Gantar, R. Hrovat, I. Kolenc, J. Therm. Anal. 36 (1990) 685.

[199] J. Macek, R. Hrovat, B. Novosel, J. Therm. Anal. 40 (1993) 335.

[200] H. Funk, A. Eichoff, G. Giesder, Omagiu Raluca Ripan, Acad. Soc. Romania, S (1966) 247.

[201] P. Ravindranathan, "Studies on metal hydrazine carboxylates: Precursors to fine particle oxide materials", Ph.D., Thesis, Indian Institute of Science, Bangalore, India (1986).

[202] P. Ravindranathan, K.C. Patil, J. Mat. Sci. Lett. 5 (1986) 221.

[203] B.N. Sivasankar, S. Govindarajan, Indian J. Chem. 33A (1994) 329.

[204] K. Kuppusamy, S. Govindarajan, Synth. React. Inorg. Met-org. Chem. 26 (1996) 225.

[205] R. Manimekalai, "Synthesis, characterization and applications of transition metal complexes of hydrazine with carboxylic acids" Ph.D., thesis, Bharathiar University, Coimbatore, India (2007).

[206] L. R. Gonsalves, S. C. Mojumdar, V. M. S. Verenkar, J Therm Anal Calorim. 104 (2011) 869.

[207] J.S. Budkuley, "Studies on hydrazine hydrate-sulphur dioxide-Transition metal ion system", Ph.D., Thesis, Indian Institute of Science, Bangalore, India (1987).

[208] P. Ray, B.K. Goswami, Z. Anorg. Allgem. Chem. 168 (1928) 329.

[209] J.S. Budkuley, K.C. Patil, Synth. React. Inorg. Met.-org. Chem. 19 (1989) 909; 21 (1991) 709.

[210] B.N. Sivasankar, S. Govindarajan, Synth. React. Inorg. Met.-Org. Chem. 24 (1994) 1583.

[211] Vogel, "A Textbook of Quantitative Inorganic Analysis", 4th Ed., Longman, UK, 1985.

[212] C.N.R.Rao, "Chemical Applications of Infrared Spectroscopy," Academic Press, New York, 1967.

[213] K. Nakamato, "Infrared spectra of Inorganic and Coordination compounds," John-Wiley and Sons, New York, 1963.